SCIENCE IN THE PRIVATE INTEREST

SCIENCE IN THE PRIVATE INTEREST

HAS THE LURE OF PROFITS CORRUPTED BIOMEDICAL RESEARCH?

Sheldon Krimsky

ROWMAN & LITTLEFIELD PUBLISHERS, INC.
Lanham • Boulder • New York • Oxford

ROWMAN & LITTLEFIELD PUBLISHERS, INC.

Published in the United States of America
by Rowman & Littlefield Publishers, Inc.
A wholly owned subsidiary of The Rowman & Littlefield Publishing Group, Inc.
4501 Forbes Boulevard, Suite 200, Lanham, Maryland 20706
www.rowmanlittlefield.com

PO Box 317
Oxford
OX2 9RU, UK

British Library Cataloguing in Publication Information Available

Library of Congress Cataloging-in-Publication Data

Krimsky, Sheldon.
Science in the Private Interest: Has the Lure of Profits Corrupted
Biomedical Research? / Sheldon Krimsky.
p. cm.
Includes bibliographical references and index.
ISBN 0-7425-1479-X
1. Medicine—Research—United States—Finance. 2. Academic-industrial collaboration—United States. 3. Biotechnology industries—United States. I. Title.
R854.U5K75 2003
362.1'07'2073—dc21

2003000225

Printed in the United States of America

∞™ The paper used in this publication meets the minimum requirements of American National Standard for Information Sciences—Permanence of Paper for Printed Library Materials, ANSI/NISO Z39.48-1992.
Manufactured in the United States of America.

Dedicated to my grandson, Benjamin Clossey.

CONTENTS

FOREWORD

The encroachment of the civic values inherent in the independence of a university by the insinuating rush of commercial values has a long history. In 1963, Clark Kerr, president of the University of California, wrote in his book, *The Uses of the University* (Harvard University Press) that "Universities have become 'bait' to be dangled in front of industry, with drawing power greater than low taxes or cheap labor." But where Kerr saw this condition as a challenge to the adapting intellect, Thorstein Veblen viewed the governance of universities by businessmen as a stifling bureaucracy over free and open inquiries by scholars. Over time, with the emergence of more corporate sector academic funding and government inducements for university–industry partnerships, with their attendant benefits and risks, the nexus of commerce with academic norms has evolved from the occasional to the frequent, and now to the customary state of affairs.

That which has been called the inevitable tide of corporate and academic partnerships and the commercialization of knowledge has, with few exceptions, hardly disturbed the leisure of the "theory class" on campus. There is little demand for disclosure or even the placement online of university–corporate contracts. The comfortable incentives that still the voices of dissent remain the routines of teaching and publishing within ever more narrow specialties. Only recently, with the specter of distant learning threatening the work product of professors, has the professoriate been attentive to colleagues like David Noble, who point out the consequences that flow from the commodification of knowledge for corporate profit.

Professor Sheldon Krimsky, the author of this impressive synthesis and critical inquiry into the conflicts arising when public science becomes entangled with private interests, is decidedly not part of the leisurely "theory class." He summarizes the theses of his book admirably in his introduction. His concluding sentence may be the most important and bears repeating: "As universities turn their scientific laboratories into commercial enterprise zones, and select faculty to realize these goals, fewer opportunities will exist in academia for public interest science—an inestimable loss to society." In chapter 11, Krimsky describes the courageous careers of three university-based public interest scientists and their contributions to an otherwise defenseless population that was the focus of their studies. Two of them were subjected to merciless pressures by commercial interests but fortunately they persisted, and now we see clearly how their work contributed to the betterment of society. We will never know, apart from anecdotal narratives, how many would-be public interest scientists, working with communities of need, have been blocked by the censorious climate of commercial supremacies on campus.

The aggressiveness of corporate science has placed academic science in an increasingly zero-sum relationship. The former comes with money to the university and prospects of personal profit to the professors. The web of influence and seduction that is woven into the fabric of academic relations is composed of many strands, with many rationalizations, but the effects are real. Academic science, with its custom of open exchange, its gift relationships, its willingness to provide expert testimony that speaks truth to power, its serendipitous curiosity and its nonproprietary legacy to the next generation of student-scientists, differs significantly from corporate science, which is ridden with trade secrets, profit-determined selection of research, and awesome political power to get its way, whether by domination or servility to its payers.

The list of neglected human needs and injustices by the paucity of free-minded scientists is legion. In areas such as tobacco, product safety, environmental pollution, workplace toxics, and the efficacy and side effects of drugs on adults and children, I have observed the difficulties of securing the participation of academic scientists for over four decades. In a climate that sanctions corporatism over individuals, the university campus has thinned the ranks of these people. The results of their work are avoidable deaths, injuries, and disease. The consequences of their decreased numbers are the reduced capacities to foresee and forestall dangerous products and technologies.

The historical stagnation of automotive technology in the area of safety, fuel efficiency, and emissions is a case in point. For many years, while the death and disease toll mounted, the perverse rewards were for academic silence, self-censorship, and distorted research priorities, which do not incur the displeasure of the corporate world or the pressures of influential alumni.

Heavy governmental contracts for military weapons research led to a major "brain drain" on campus. Civilian needs received little research attention.

Expectation levels declined as well for the university's multiple expressions of learned traditions, independence, and public service. It was not surprising that students came to their universities with similarly low expectations, viewing their years there as opportunities for vocational training more than liberal arts education, for commercially viable skills without literate citizen skills for negotiating and building democracy as a central problem-solving process. That we live in a society with far more problems than it deserves and far more solutions than it applies is not unrelated to the erosions and conflicts of interest described and analyzed by Krimsky in this searching and honest book.

In recent years I have been asking a number of university presidents whether they have a comprehensive written policy toward demarking the line beyond which corporate commercialism cannot cross. Their replies indicate that their policy is ad hoc; ad hoc then becomes a moveable set of expediencies without a well-conceived public philosophy regarding the content of academic freedom, public service, and intellectual independence.

In his small insightful book *In the University Tradition* (Yale University Press, 1957), Whitney Griswold, the late president of Yale University, decried the "little comprehension of the power of the liberal arts in American society" and attributed this condition to our "incurious and inarticulate" state "concerning its own political and social philosophy." Our universities are exhibit one for not keeping "fresh and vital their meaning."

A sustained movement of faculty and students is needed to defend and expand the independence of the university, its academic freedoms, and its civic values from the growing corporate state. The relevant frameworks are the very ideals of minimum tuitions for maximum access, of educating for future civic leadership, of linking knowledge to its application for human betterment, of having an intellectual life on campus beyond narrow specialties, of bringing together, in the words of former Cornell University president Frank H. T. Rhodes, students with "a community of more senior scholars to reflect on the great issues of life and to confront the over-arching challenges to society."

Yes! The university itself should be a proper subject of study by its students as part of the curriculum, year after year. As an intellectual subject that combines all rungs of the abstraction ladder, it has few peers in its immediacy, its pertinence and its provocative effect on the life of the mind in action. Reading Krimsky will give those inside and outside the university and college worlds a gripping sense of how large are the stakes and how glorious can be the benefits.

Ralph Nader
Post 19367
Washington, DC 20036

ACKNOWLEDGMENTS

I would like to thank the Rockefeller Foundation for providing financial support for this book, and I would also like to thank Tufts University for providing a sabbatical leave that afforded me the unencumbered time to begin this inquiry.

The following people were gracious enough to either give me feedback on earlier versions of the work (or parts thereof), contribute source materials, or agree to be interviewed: Carol Ahearn, Gerard Clark, Luz Claudio, Barry Commoner, John DiBiaggio, Ralph Knowles, Claire Nader, Herbert Needleman, Stuart Nightingale, L. S. Rothenberg.

I am also indebted to the many fine investigative journalists and science reporters who have contributed numerous columns of print on issues discussed in this book. Their stories have sensitized the public to the problems of research ethics and scientific integrity. The following is but a short list of journalists whose work I have read and benefited from (some of their affiliations, however, may well have changed over the two decades that I have followed their stories): Chris Adams, *Wall Street Journal;* Goldi Blumenstyk, *Chronicle of Higher Education;* Georgette Braun, *Rockford Register Star;* Matt Carroll, *Boston Globe;* Dennis Cauchon, *USA Today;* Alice Dembner, *Boston Globe;* Kurt Eichenwald, *New York Times;* Peter Gorner, *Chicago Tribune;* Peter G. Gosselin, *Boston Globe*, *Los Angeles Times;* Jeff Gottlieb, *Los Angeles Times;* Lila Guterman, *Chronicle of Higher Education;* David Heath, *Seattle Times;* Andrew Julian, *Hartford Courant;* Dan Kane, *News & Observer;* Mathew Kaufman, *Hartford Courant;* Ralph T. King, *Wall Street Journal;* Will Lepkowski, *Chemical & Engineering News;* Maura Lerner, *Star Tribune;* Kitta MacPherson, *Star-Ledger;* Terrence Monmaney, *Los Angeles Times;* Tinker Ready, freelance; Joe Rigert, *Star Tribune;* Steve Riley, *News & Observer;*

Edward R. Silverman, *Star-Ledger;* Sheryl Gay Stolberg, *New York Times;* Elyse Tanouye, *Wall Street Journal;* Jennifer Washburn, *Washington Post;* Rick Weiss, *Washington Post;* David Wickert, *News Tribune;* David Willman, *Los Angeles Times;* Duff Wilson, *Seattle Times;* Trish Wilson, *News & Observer;* Mitchell Zuckoff, *Boston Globe.*

1

INTRODUCTION

Modern governments have become inordinantly dependent on expert knowledge for so many of their decisions that, as individuals, we rarely have the luxury of questioning whether the expertise is trustworthy. An investigative story will periodically call into question the science behind a study, or it may even highlight disagreements among scientists over the potential hazards of a product. But the majority of cases involving scientific expertise is never aired in public venues, either because they are under gag orders while in litigation or because they fall below the radar screens of the media. I have been following the use of expertise in science for several decades, and I am convinced that there have been significant changes in the culture, norms, and values of academic science during this period. My focus has been largely in biomedical science, but I have reason to believe that the transformations taking place in this field can be found in other fields as well. We get a glimpse of these changes from periodic headlines in popular newspapers and magazines: "New corporations could turn academics into tycoons." "Public corporations enrich drug makers, scientists." "Biomedical results are often withheld." "Surge in corporate cash taints integrity of academic science."

These and other stories are beginning to paint a rather unseemly picture of the landscape of the American academies of higher education. In such a picture, university science becomes entangled with entrepreneurship; knowledge is pursued for its monetary value; and expertise with a point of view can be purchased. The new image of the successful and admired scientist is no longer modeled on public scientists such as Linus Pauling and George Wald, who devoted themselves to pure science and to questioning science's role in the betterment of society. Rather, the successful scientist today is the person who can make contributions to the advancement of knowledge while concomitantly participating in the conversion of the new knowledge into marketable products. As for the "betterment of society," such a notion now becomes captured by the phrase "knowledge transfer." It is said today that the scientists who can turn ideas into profits are the ones who are contributing to a better world. Conversely, those scientists who stop at merely generating fundamental knowledge are the ones who allow their discoveries to remain unrealized possibilities for the marketplace.

Why should this paradigm be problematic? Isn't this scenario the means through which technological societies have created new wealth and improved the quality of lives of their citizenry? No one can doubt the role that science and technology have played in contributing both to the wealth and power of nations, but the question at hand is whether universities should be turned into the instruments of wealth rather than protected enclaves whose primary roles are as sources of enlightenment. Is there any contradiction in these two goals?

When the pursuit of knowledge is turned in the direction of commerce, a suspicion inevitably accompanies it. Will the thirst for financial success bias the objective assessment of scientific truth? Will the public perception of the scientist become impregnated with skepticism and mistrust? Is there any distinction between academic science and industry science when the demarcation between academia and industry becomes blurred?

The last quarter of the twentieth century was a period when many of the country's leading universities began to redefine their mission. University presidents and trustees began to ask themselves the following questions: How does a modern research university stay competitive when federal budgets for science and applied medical research are not secure? How can a university diversify its funding sources to protect itself from rapid fluctuations or downturns in endowment investments, in government-sponsored research, and alumni giving? How do universities create new wealth for building the modern fiber-optic campus with state of the art research facilities in frontier areas of science and engineering when there are limits on tuition-generated income? What can universities do to protect their faculty from leaving for more lucrative jobs in the private sector or from being hired by competitor institutions?

Around the same time these discussions were taking place, the federal budget became a target for reform. An idea had caught the interest of Congress. The idea was that universities should learn how to generate wealth by partnering with the private sector and by selling their knowledge and patentable discoveries. According to this view, universities are sitting on unrealized wealth in the intellectual work of their faculty. The new operative terms for success are therefore university–industry collaborations, technology transfer, and intellectual property. Once the incentives were in place, these new relationships were crowned as a triple-win strategy.

The theory was that universities and their entrepreneurial faculty would win by deriving new sources of wealth from licensing discoveries and from developing equity interests in faculty companies. The private sector would gain by signing lucrative research contracts and licensing agreements with universities, which would result in high-value products such as drugs and new technologies. The public would eventually win by having new products and therapies that might not have been developed were it not for the collaborations. The fact that this scenario appears to create so many winners makes it difficult for one to think about the liabilities—and there are liabilities to the new aggressive commercialization of universities and nonprofit research centers. These liabilities are subtle and more diffuse than the benefits. The adverse consequences manifest themselves over a long time period, and they rarely produce a dramatic outcome. Nonetheless, the transformation in American universities as a result of the new ethos of commercialism will have pernicious effects.

So much has been written in praise of the new commercial constructs of universities that questions about the impact of these relationships on the integrity of the academic institutions are rarely asked. When they are addressed, they focus almost exclusively on how universities can accommodate to the new norms of academic science. In other words, rarely, if ever, do they discuss these changes in the context of the social value universities play in nurturing public interest activities, where professors actively exercise their rights of academic freedom toward social ends.

The primary argument of this book is that the new ethos of academic commercialism, operating at all levels of research institutions, is largely viewed as a proper, favorable tradeoff of values, where conflicts of interest are manageable and impossible to eliminate and where the basic integrity of the university can be protected. My contention is that the most significant loss in permitting academic scientists to pursue technology transfer, to establish new companies in partnership with the university, to exploit intellectual property of scientific knowledge is that it turns the university into a different type of institution. The greatest losses are not to the academic professions or to the scholarly publications, but rather to the social role played by universities in American life. Protecting the integrity of America's research institutions is like protecting a unique natural resource such as the Grand Canyon from being exploited for its unrealized wealth in precious metals. Universities are more than the wellsprings of wisdom. They are the arenas through which men and women of commitment can speak truth to power on behalf of the betterment of society. The patchwork effort to reconcile the values of academic science with the values of business enterprises misses the hidden loss to society as a result of a hybrid institution. When universities and federally supported nonprofit research institutions are turned into private enterprise zones, they lose their status as independent and disinterested centers of learning. Moreover, they no longer provide as favorable an environment for nurturing public interest science—those multifarious contributions of academic scientists to all levels of government and across vulnerable communities that address social and environmental problems.

While writing this book, I have had to confront a number of tendentious assertions about the university as an institution and about science. These will be discussed throughout the volume, and they include the following:

- Financial conflicts of interest are not the only conflicts found among university scientists, and they are hardly the most important.
- Science has its own system of ethics based on the pursuit of truth, and it does not require accountability outside of its own institutions and the system of peer review.
- Conflicts of interest and university–industry partnerships will not affect the quality of science because no direct relationship exists between having a conflict of interest and research bias.
- There is nothing new or unique about university–industry partnerships. Collaborations between academia and industry have always been taking place.
- It is not possible to simultaneously support academic freedom while preventing conflicts of interest at universities.
- The rise of conflicts of interest is a result of universities' and their faculty's making a free and independent choice to trade off some standards of research integrity for financial gain.
- Policies promoting disclosure of conflicts of interest will resolve the problem of public trust in science.

In this book, I share with the reader my personal explorations, investigations, and discoveries regarding the increasing influence of commercialism in academic science and biomedical research. The mix of science and commerce continues to erode the ethical standards of research and diminish public confidence in its results. I have learned a great deal from the many reports of our finest investigative journalists, from personal interviews, and from scores of books and studies that have tried to make sense out of the rising tide of scientific entrepreneurship. This book examines the factors that helped bring about the new incentive structure in universities for making knowledge production into a business. Fears about declining federal support for research awakened university administrators to the need for new sources of revenue. Business school managers took on the task of making universities function more efficiently. Campuses around the country began to privatize many of their services at a time when tuitions were rising faster than the cost of living. Meanwhile, federal policies created tax and funding incentives for the establishment of university–business partnerships. For the first time since the period following World War II, federal research policy became part of a new agenda for industrializing the American economy. Each of the chapters in this book provides a window into the changes and dilemmas brought about by those changes that commercialization of science and the commodification of knowledge have introduced into our institutions of higher learning.

Chapter 2 presents a series of cases that explore the influence of private interests on the scientific agenda. Although these cases are emblematic of the changing mores that are taking hold in the scientific culture, they also highlight the challenges faced by institutions that are seeking to protect the integrity of the research enterprise.

Chapter 3 examines the conditions that brought about the new relationship between universities and the for-profit sector. A generation of scientists conscientiously departed from the traditions of their mentors and created "academic enterprise zones," and they did so for specific reasons. It is much too simplistic to say, "Scientists made discoveries, saw opportunities to become rich, and made a rational, economic choice to add venture capitalism to their résumé." A host of incentives were put into place by governmental sectors for creating closer linkages between universities and industry. Acting on these incentives, many universities began investing in venture capital companies, and they soon become partners with members of their faculty. These arrangements have raised a new ethical quandary—namely, how to deal with institutional conflicts of interest. For example, suppose a faculty member publishes a study that is critical of a product in which the university has an investment. Will the university be inclined to punish one of its own, when the financial stakes become high enough? Will the right to publish and quickly disseminate research findings be tempered by institutional earnings?

In chapter 4, I discuss the legal and economic rationale for patenting living organisms and components of living things, and I explore its impact on commercializing science. For instance, how did we arrive at a point where it is acceptable to take out patents on the genetic code? I shall trace the logic of the Supreme Court decision in 1980 that approved the first patent for a life form to later decisions of the U.S. Patent and Trademark Office that approved patents for animals, genes, and even subgenomic bits of DNA. I also discuss the implications of gene patents as incentives (or disincentives) for innovation in biomedical research.

In chapter 5, I look back into the early twentieth century, when the image of science was being recast from the ideals rooted in the Enlightenment. The scientist was characterized as a person of unusual intellectual curiosity who embraced a code of honor—one that was unspoken, but nonetheless universally shared. According to this code, the practice of science required the universal acceptance of certain ethical standards. In other words, there was no need to keep watch over scientists. The collective good of science was linked to the adoption of certain norms of behavior, such as those framed by the sociologist Robert Merton. The chapter reexamines the Mertonian norms in light of the current conditions of scientific practice.

Chapter 6 discusses the role that scientists play on federal advisory committees, and it also examines the safeguards that are supposed to apply to prevent conflicts of interest from creating mistrust in the process. The federal laws governing advisory committees are explicit about conflicts of interest; nevertheless, the public is still ill-informed about how effectively (or ineffectively) the scientific advisory process adopted by government protects us against vested interests.

Chapter 7 examines the trend of dual affiliations held by university scientists who develop consulting firms and start venture capital companies while holding down full-time academic jobs. It discusses the ethics of health scientists, well paid by pharmaceutical and medical-device companies, who promote products through ghostwritten articles and who undertake research that consciously or unconsciously carries a sponsor's bias.

Chapter 8 explores the meaning of conflict of interest at a time when so many academic scientists have affiliations with nonprofit as well as profit-making institutions. The rise in multivested science has called into question the traditional definitions of conflicts of interest and the role those definitions play in setting ethical standards for science. I question whether it is possible to preserve the traditional meaning of "conflict of interest" when the practices that ordinarily exemplify the use of the term have become so pervasive that they have lost their pejorative connotation. I show how conflict of interest in science has become the norm of behavior, rather than the exception.

Chapter 9 poses the following question: Do conflicts of interest in science affect the outcome of research studies? If they don't, then why all the fuss? Is it simply a matter of *appearing* unbiased? But what if there were evidence that demonstrates that "possessing financial interests" can be a source of scientific bias? I describe one study that measured the amount of financial interest in scientific publications, and I describe others that found a relationship between privately funded science and the outcome of research.

Chapter 10 summarizes what we know about conflicts of interest in scientific publications, including the policies and the practices of journal editors who manage those conflicts. How many scientific journals have conflict-of-interest policies? Are the policies effective? Do journal editors disclose the financial interests of authors, or do they protect such information as proprietary?

Chapter 11 discusses the distinctive features of universities, and it examines the concepts of public-centered and private-centered science. Scientists who apply their expertise to address public problems are defined as "public-centered," while those who turn their expertise toward commercial interests are defined as "private-centered." I explore the historical paths, the challenges, tribulations, and personal gratification of three notable scientists, whose careers exemplify the ideal of public-interest science. This chapter argues that the new commercialization of university science will eventually deplete the public-interest roles of scientists who have contributed so much to society.

Chapter 12 addresses a variety of ethical and institutional responses to the new commercialism of academic science from universities, journals, government, and professional societies. The rush toward privatization has prompted agencies of government and universities to search for new equilibria points. Their mission is to balance the interests of university scientists so that they can innovate for profit while preserving the unique values of the American university as a source of dependable and trustworthy knowledge and where academic freedom is protected. The chapter reflects on whether these initiatives on conflicts of interest and academic entrepreneurship are sufficient to bring a new moral sensibility to science.

Finally, chapter 13 questions the value of academic freedom in the new entrepreneurial university. It frames a set of principles for recalibrating the moral boundaries of scientists and medical researchers, and it illustrates how those principles can be applied to prevent conflicts of interest.

The central argument of the book that ties the chapters together may be summarized as follows: Public policies and legal decisions have created new incentives for universities, their faculty, and publicly supported nonprofit research institutes to commercialize scientific and medical research and to develop partnerships with for-profit companies. The new academic–industry and nonprofit–for-profit liaisons have led to changes in the ethical norms of scientific and medical researchers. The consequences are that secrecy has replaced openness; privatization of knowledge has replaced communitarian values; and commodification of discovery has replaced the idea that university-generated knowledge is a free good, a part of the social commons. The rapid growth of entrepreneurship in universities has resulted in an unprecedented rise in conflicts of interest, specifically in areas sensitive to public concern. Conflicts of interest among scientists has been linked to research bias as well as the loss of a socially valuable ethical norm—disinterestedness—among academic researchers. As universities turn their scientific laboratories into commercial enterprise zones and as they select their faculty to realize these goals, fewer opportunities will exist in academia for public-interest science—an inestimable loss to society.

2

TALES OF THE UNHOLY ALLIANCE

Several observations can be made about the conflicting roles of university scientists. First, there are vastly more conflicts of interest than meets the eye. What we hear about is the proverbial tip of the iceberg with regard to conflict-of-interest cases. The great majority of cases remain undisclosed. Second, the public is more likely to learn and show concern about the conflicting interests of a researcher when something goes amiss, such as an adverse drug event during clinical trials or when the scientist becomes embroiled in a public controversy. Scientists with conflicts of interest have been known to follow a policy likened to the military's view of gay soldiers—"don't ask, don't tell." Third, the most common source of public knowledge that a scientist is involved in a conflict of interest comes from investigative journalists. Neither government agencies nor universities are likely to report to the public the conflicting interests of their researchers without pressures from the press. Fourth, rarely is anyone's position threatened from having vested interests that contribute to or give the appearance of bias. Scientific conflicts of interest are taken much less seriously than allegations of fraud or misconduct, and in many universities, they are accepted as the norm for researchers. The operative term is "managing" rather than "avoiding" or "preventing" conflicts of interest.

As a result of these factors, print journalism plays a critical role in bringing the more egregious conflict-of-interest cases to the attention of the public and the policy makers. The best investigative stories may send out a cautionary signal, but rarely do they provide the impetus for any significant changes in policy. However, investigative journalism does have an accumulative effect. Somehow, perhaps because of their own culture, where conflicts of interest are taken very seriously, print journalists have not acquiesced to the normalization of "conflicts of interest" among scientific investigators. Each case that is reported gives other journalists a sense that they share in a collective mission to make transparent the hidden biases that permeate certain fields of scientific research.

The following cases have earned a reputation as reference cases for the types of conflicting interests that are occurring at an ever greater frequency within the academy and within the research centers of government. We should keep in mind, however, that conflict of interest among scientists is not a uniquely American phenomenon. The idea of university–industry partnerships has been exported to European countries and Asia, which are beginning to face similar problems of trust within their scientific institutions. For example, at Swedish universities, which are state owned, private corporations are financing an increasing proportion of the professorships. At the prestigious Karolinska Institute, where an estimated one-third of the professorships are financed by private companies, the pharmaceutical giant Astra-Zeneca has exclusive property rights to all results of one professor's research in neurology, including results from research that is partially financed by the Swedish National Medical Research Council. That professor is salaried by the company.[1]

A Polish medical scientist summarized the state of affairs as follows: "Privatization and commercialization are threatening the objectivity of clinical research and the availability of health care because uncontrolled market mechanisms focused on profit are nurturing conflict of interest that generate bias and unreliability into research and medicine."[2] An Italian editor of an international medical journal wrote: "Members of corporate driven special interest groups, in virtue of their financial power and close ties with other members of the group, often get leading roles in editing medical journals and in advising non-profit research organizations. They act as reviewers and consultants with the task of systematically preventing dissemination of data which may be in conflict with their special interests."[3]

CASE 1: HARVARD OPHTHALMOLOGIST MAKES MONEY OFF OF HIS COMPANY'S UNPROVEN DRUG

The *Boston Globe* banner headline, along with a photograph of a Harvard clinician, read, "Flawed Study Helps Doctors Profit on Drug." The investigative report by journalist Peter Gosselin ran on October 19, 1988. Gosselin traced a woman from Drescher, Pennsylvania, who received treatment at the Massachusetts Eye and Ear Infirmary beginning in the late fall of 1985 for a condition called "dry eye."

The physician who supervised her treatment was Dr. Schaefer Tseng, an ophthalmologist who received his undergraduate and medical degrees from the National Taiwan University and a Ph.D. from the University of California, San Francisco. Tseng was on a two-year fellowship at Harvard, where he was doing research at Massachusetts Eye and Ear (Mass Eye and Ear) Hospital.

At about the time the woman from Drescher came to Boston, Tseng was forming a clinical trial group of about 300 individuals. He accepted the young woman into the drug study. The patients in the experimental group, which included the woman, received a vitamin A ointment for their "dry eye" condition. None of the participants in the clinical trial was aware that the drug they were taking was manufactured by a new company called Spectra Pharmaceutical Services (or Spectra for short). The company was founded by Tseng and several of his colleagues. The *Boston Globe* investigation learned that Tseng's supervisor at Mass Eye and Ear, a teaching hospital for Harvard, had equity interests in Spectra. So did Edward Maumenee, a professor emeritus of ophthalmology at Johns Hopkins Medical School, under whom Tseng had studied. He also chaired Spectra's board of directors.

During this period in the 1980s, it was becoming more commonplace for academic physicians to pair up with or form their own companies. Pharmaceutical companies submit data to the Food and Drug Administration on the safety and efficacy of new drug applications. They depend on university physicians to carry out clinical trials that generate the supporting data for these applications. Rarely does the government undertake its own studies. Research physicians began to see the financial opportunities of venture capital start-up companies, which do all the research and development work on specialized drugs until either they issue stock in a public offering or are bought up by a large pharmaceutical company.

In 1980, Arnold Relman, then editor of the *New England Journal of Medicine*, wrote an editorial that sounded the alarm about clinicians who are involved in for-profit ventures. He noted that involvements of physicians with drug or health-product companies represent a direct conflict of interest with their professional responsibility to their patients. Relman viewed the commingling of clinical practice with commerce as morally unsupportable: "What I am suggesting is that the medical profession would be in a stronger position, and its voice would carry more moral authority with the public and the government, if it adopted the principle that practicing physicians should derive no financial benefit from the health-care market except from their own professional services."[4] Twenty years later an editorial in *The Lancet* spoke soberly about the changing culture of biomedical science: "Today's universities are increasingly encouraging their scientists and doctors to be entrepreneurs and to commercialize their intellectual property. However, the collaboration between industry and academia or the combining of private and public interest can easily end in tears."[5]

There are three noteworthy aspects about the Tseng affair. First, it was not the practice during those years to tell patients that the experimental drug being tested in the clinical trial was produced by a company that was established by their physician. That scenario was certainly true in this case. How many people might refuse to participate in a clinical trial if they were privy to this information? Some patients might feel that their physician, who had a direct financial investment in the drug, could not provide an objective assessment of its effectiveness. Would they be too eager to see the drug's benefits and not its risks? Would they err on the side of investor interests rather than patient interests?

Second, changes were made in the protocols of the clinical trial, such as extending the use of the drug on other patients and changing the dose, without receiving permission from the human subjects committee, also known as the Institutional Review Board (IRB) of the hospital. One cannot draw the conclusion that violations in the IRB rules are more likely to occur when scientists have financial stakes in the outcome; however, the conflicts of interest cast a cloud of suspicion over the clinical researchers that places IRB violations into a more serious context.

Both Mass Eye and Ear Hospital and the Harvard Medical School did independent investigations of Tseng's activities and according to the *Boston Globe* found "violations of rules governing patient testing, conflicts of interest, research quality and supervision."[6]

Third, the physicians of Spectra Services seemed to act strategically in their use of the media to raise investor confidence on the basis of past research findings, while selectively hiding new results that were not nearly as favorable to the company. The Food and Drug Administration has legal authority to punish drug companies that selectively present data on their products or mislead physicians and the general public about the properties of the drug, including effectiveness and risks. The scientists at Spectra first thought about which results they would reveal to the media and to the investment community; they then thought about when to reveal these results so that they could maximize the value of the public stock offering. The *Boston Globe* reported that Tseng and his family made at least $1 million in 1988 through stock sales of Spectra.

Tseng was cited by institutional officials for adding new patients to his clinical trials without notifying the hospital's human subjects committee and for violating hospital and university ethics guidelines by administering an experimental drug without disclosing its name. By the fall of 1988, Tseng and his partners published the results of the vitamin A cream in which they claimed that it may help in treating some rare eye disorders, whereas for "dry eye" (the reason many patients sought help), the placebo may be more effective.

Why were physicians at Mass Eye and Ear permitted to dispense a drug produced by a company in which they had equity interests? Why were medical researchers allowed to administer a drug in a clinical trial without fully informing the patient about the nature of the drug and their relationship to it? Why were there selective media releases of the results of trials before peer-reviewed articles were published?

The Tseng affair prompted Harvard and other medical schools to reexamine their ethical standards of research. How can universities claim their objectivity in the search for truth when they are willing to accept such blatant conditions of conflict of interest? No publicized evidence exists that claims Tseng and his colleagues in this case falsified data or gave biased interpretations of any results. What they did was violate ethical guidelines on the conduct of human experiments and release information selectively when it advanced the stock interests of Spectra Pharmaceutical Services. Interestingly, those latter actions did not produce an investigation by the Securities and Exchange Commission for any violations of securities law.

Charges against Tseng and his business associate Kenneth Kenyen were brought to the State Medical Board of Massachusetts. The hearing officer found that Tseng violated hospital policy and the clinical study protocol (a finding of the magistrate assigned to the case), yet the charges filed with the State Medical Board were dropped by the magistrate. The putative reason for dropping the charges was the lack of evidence that Tseng committed fraud or was involved in unethical behavior, even though it was reported that he had administered drops containing a substance that had not been approved by the human subjects committee for experimental use on humans.

Medical scientists associated with prestigious universities are uniquely situated to set up companies that selectively release scientific findings to the media before peer review for the purpose of creating a stock demand that can bring quick riches to the principal equity holders. The *Boston Globe*'s Gosselin discovered that "their plan was what is known in the brokerage industry as a 'rich' deal. The insiders, including Maumenee, Tseng, and Kenyon, would keep more than three-quarters of the shares for themselves, but put in less than 3 percent of money raised by the stock sale. The investing public would put in the other 97 percent for the remaining quarter of the stocks."[7]

Traders in public securities depend on the trustworthiness of biomedical scientists. Without the names of prestigious scientists on the boards of new companies, who would invest? Scientists leverage the prestige of their universities to attract public investors while university officials are kept in the dark, unless they are partners in the venture company. Scientist-entrepreneurs are insiders to both the science and the company that stands to gain from certain published or reported results. They can buy and sell stocks based on information they provide to the media and the investment community. From 1986 to 1988, the stock of Spectra reached a high of eight-and-a-quarter and a low of three-eighths. However, scientist equity holders in the company were able to make windfall profits because they were in control of information that the public did not have. Rarely, if ever, are scientists convicted of insider trading—a practice that is difficult to avoid when the roles of scientists include the pursuit of knowledge, the management of information transfer, and the maximization of the stock portfolios.

It is difficult to gauge the total amount of corporate funds that is channeled into universities annually because no centralized data bank exists for this type of information. In addition to the grants and contracts that are registered with the university, researchers also receive unregistered gifts as well as consultantcies, honoraria, travel grants, and expert witness fees. Considering only registered grants and contracts, which are tracked, the amount of corporate research funds (in constant dollars) invested in academia over two decades rose from $264 million in 1980 to over $2.3 billion in 2000.[8] As federal dollars fund about $17.5 billion in university research, corporate contributions are estimated to have reached 13 percent of government research contributions during the last decade. Some corporate sectors are more highly dependent on university scientists, which is the exact case with the pharmaceutical industry. Drug companies must have credible studies to gain drug approval from the Food and Drug Administration. Even after a drug has been approved, a company faces challenges from competitors about the comparative safety and efficacy of its drugs. As a drug patent nears maturity, companies depend on biochemists to modify the formulation just enough to gain a new patent without compromising the efficacy. The next case illustrates the problems that can arise when a scientist signs a research contract with a pharmaceutical company and does not read the small print.

CASE 2: SCIENTIST SIGNS A RESEARCH AGREEMENT WITH A DRUG COMPANY AND LOSES CONTROL OVER THE SCIENCE

In 1986, a scientific letter appeared in a specialized journal in the field of clinical pharmacology stating that not all products on the market treat hypothyroidism equivalently. The authors of the letter cited two brand-name preparations as the drugs of choice, one of which was Synthroid.[9] Betty J. Dong, a doctor of pharmacology who had an appointment at the University of California at San Francisco (UCSF), was one of the authors of the letter. Dong was hired as an assistant professor at UCSF in 1973 after she had completed her residency there in clinical pharmacy. Naturally, the company that manufactured the drug Synthroid, Flint Laboratories, could not have been more pleased in seeing such a letter confirm what it already believed—namely, that its product was a leader in the field of synthetic thyroid drugs and that it was superior to the generics.

In the United States, approximately eight million people are afflicted with hypothyroidism, a disease associated with a deficiency of thyroid hormone. Synthroid, the first of a generation of synthetic thyroid drugs, dominated a market said to be valued at about $600 million per year. When Synthroid was first approved by the FDA, the agency did not have standards for the bioequivalency of drugs. More recently, the FDA has established criteria for testing whether two drugs are bioequivalent. Test results on bioequivalency are used by the FDA to guide clinicians on the choice of drugs and to determine whether the drug companies are providing truthful advertisements for their products.

In 1987, Synthroid was facing competition from cheaper generic drugs, which had the potential to erode its 85 percent market share. Flint Laboratories felt it was a good time to establish the superiority of its product against the generics. In a twenty-one-page contract it signed in May 1988 with Betty Dong and officials at UCSF, the company committed to funding a six-month human study to assess the bioequivalency of the major drugs (one other brand name, Levoxyl, and two generics) for hypothyroidism. The $256,000 contract contained details about the experimental design and the analysis of the data. But it also contained a clause, one that is sometimes referred to as a "restrictive covenant," which stated: "All information contained in this protocol is confidential and is to be used by the investigator only for the conduct of this study. Data obtained by the investigator while carrying out this study is also considered confidential and is not to be published or otherwise released without written consent from Flint Laboratories."[10]

Two questions come to mind at this point. Why did UCSF allow Dong to sign the restrictive covenant, which would limit her control over her study? Was Dong aware of the restrictions placed on her research? At the time Dong signed the contract with Flint Laboratories, her university did not have a policy on restrictive covenants. A year later in 1989, a revised UCSF policy read: "The University will undertake research or studies only if the scientific results can be published or otherwise promptly disseminated."[11] The university left it up to the investigator to use good judgment in working out the contractual arrangements for interpreting what it means for work to be "promptly disseminated." Dong subsequently expressed some concerns about the restrictive language in the contract she received from Flint, but at the time she convinced herself that it was not unusual and would not impede her publication of results. "I wasn't enthusiastic about signing this contract, but the company assured me I would not need to worry on its bearing on the publication of the study. Colleagues that I consulted also noted that such contracts were common and not to worry. That was before the change of power."[12]

Dong is referring to the fact that Flint Laboratories was sold to Boots Pharmaceuticals, which was itself merged into Knoll Pharmaceuticals (a subsidiary of BASF AG) between March and April 1995 for $1.4 billion. During the period these corporate buyouts were taking place, the value of Synthroid's market share played a role in the purchase prices, which could explain the company's tough stance against publication of data that could diminish Synthroid's value.

When Boots Pharmaceuticals took over the rights of Synthroid, the company paid close attention to Dong's bioequivalency study by making periodic site visits to her laboratory. In one of those visits around January 1989, representatives from Boots requested preliminary results of a parallel study (not involving human subjects) of the synthetic thyroid tablets. But this information, once released, would require breaking the code of the double-blind study. (A double-blind study is one in which neither the subjects nor the researchers know which subjects received the drug and which received the placebos.) A double-blind study is the gold standard in research for minimizing both investigator and patient bias.

The point of this particular double-blind study was that the twenty-two subjects did not know the order of the drugs administered to them—although they did know that they were getting a dose of each of the four drugs. In addition, the researchers who measured their thyroid levels and took their biological measurements did not know which drug preceded those measurements. Dong and her colleagues refused to comply with Boots' requests on the grounds that it would weaken the methodology and violate the protocols of the study.

By the end of 1990, Dong completed the study and sent the study results to Boots. The preliminary findings indicated that the four tested drugs were bioequivalent. Based on this information, Boots notified UCSF administrators, including the chancellor and several department heads, and informed them that the Dong study was flawed. UCSF initiated two separate investigations of Dong's study. One investigation, conducted by a pharmacist and concluded in 1992, found that the only flaws in her study were "minor and easily correctable." A second inquiry, undertaken by the chair of UCSF's Pharmaceutical Sciences Department, found that Dong's study was rigorous and that Boots' criticisms were "deceptive and self-serving."[13]

In April 1994, Dong submitted the manuscript containing the results of the study to the *Journal of the American Medical Association* (*JAMA*). It was peer reviewed by five individuals, revised by the authors, and resubmitted. The paper was accepted for publication in *JAMA* on November 1994. One of the implications of the Dong study was that physicians could prescribe the less expensive versions of the thyroid drug and save consumers an estimated $365 million annually. This figure was of interest to health consumer groups and insurers, including the new generation of health maintenance organizations (HMOs), the prepaid medical insurers who had structural incentives to manage health care costs. Both the savings and the bioequivalency results of the study were also of interest to FDA since this agency monitored drug advertisements for false or misleading claims.

Knoll began exerting pressure on Dong and her associates about their analysis of the data. Francis Greenspan, one of her coauthors, reported that "Knoll pharmaceuticals told us that if we published the study and they lost money as a result . . . UCSF and the researchers would be liable for those losses."[14] The company claimed that its formulation of the synthetic thyroid hormone, levothyroxine, was superior to that of its competitors and that the Dong study failed to show the bio-equivalance by not analyzing the effect of the drugs on thyroid-stimulating hormone (TSH). "The Dong study," it said, "was neither an adequate nor well controlled study nor did it conclusively establish that complete interchageability exists for levothyroxine products."[15] The threat of a personal lawsuit against Dong loomed heavily on her mind. She consulted with USCF attorneys. The university lawyers told Dong and her six associates that based on the contract they signed (and that the university approved), they would have to defend themselves in court without help from UCSF. When confronted with that contingency, two weeks before the article was scheduled to be published in *JAMA*, Dong withdrew the paper.

Meanwhile, scientists at Boots/Knoll, who had been monitoring the work of Dong and her colleagues, published their own interpretation of the data collected by her research team in a sixteen-page article that did not acknowledge the work of the USCF scientists. The article, which reached the opposite conclusions of the Dong paper, appeared in a new journal to which the Boots/Knoll scientist was an associate editor.[16] The editors of *JAMA* received a letter from a scientist from Boots/Knoll outlining their criticisms of the Dong study and suggesting to the journal that they should be concerned about publishing it. The company is also said to have hired investigators to look into possible conflicts of interest of USCF scientists.

The public learned about the issue of the Dong study when the *Wall Street Journal* (*WSJ*) published an investigative report on April 25, 1996. Several months after the *WSJ* report, the FDA notified Knoll that it had violated provisions of the Federal Food, Drug, and Cosmetic Act by misbranding Synthroid. Discussions between Knoll, the University of California/USCF and

Dong continued in an attempt to resolve the issue of publication. Both the new publicity and the FDA involvement may have prompted Knoll to find a mutually acceptable resolution to the impasse over publication. By November 1996, Knoll agreed not to block the publication of the bioequivalency study of Dong and her colleagues, and publication of the paper appeared in the April 1997 issue of *JAMA*.

The attorneys general of thirty-seven states filed suit with Knoll Pharmaceutical based on the claim that the company withheld information from the FDA and distributed misleading information about its product. While Knoll continued to assert that its formulation of levothyroxine was superior, the company agreed to pay the states $41.8 million as a settlement of the suit. A class action suit was also filed against Knoll claiming that consumers were overcharged as a result of physicians' prescribing Synthroid over other formulations, a decision made on the basis of misleading information. The company agreed to pay up to $135 million to Synthroid users. The settlement was eventually reduced to $98 million since fewer than five million (out of a total of eight million) drug users signed on to the class action suit. Other problems arose when a federal judge did not approve the settlement between Synthroid users and the manufacturer because of high lawyer's fees.[17]

This case illustrates the critical importance for universities to put in place policies that oppose restrictive covenants in research contracts. University lawyers should scrutinize the language of any corporate research contract offered to an employee at their institution. Contract language that establishes sponsor control over publication or data can create irreconcilable divisions between the freedom to publish and the sponsor's legal rights over the data and interpretation of results.

The next case illustrates the inadequacy of federal standards to prevent conflicts of interest among government scientists who hold positions of influence in public health decisions.

CASE 3: GOVERNMENT SCIENTIST OVERSEES DRUG STUDY WHILE CONSULTING FOR DRUG COMPANIES

We have become all too familiar with stories of university scientists who play dual roles: the disinterested researcher pursuing knowledge for its own sake and the scientist entrepreneur seeking to profit from discovery. Less common are cases where government scientists are in conflicting roles, such as when consulting for companies while performing publicly funded studies in federal facilities. The reason why we do not expect to hear about such conflicts of interest is that the government's conflict of interest rules have been in effect longer than those in academia and that federal rules, when broken, can carry harsher penalties.

The case of the drug Rezulin, which experienced an unusually rapid entry and removal from the consumer market, brings to light some of the problems with the patchwork of conflict-of-interest policies imbricated within the federal government. After researching this case, I used a reasonable interpretation of the current government policies on conflict of interest to question whether it violated those policies. I also sought to answer how serious the transgressions would have to be for there to be violation in the event that this case complied legally.

We begin our story on July 31, 1996, when the Warner-Lambert Company submitted a new drug application (or an "NDA," as the insiders call it) to the Food and Drug Administration. The drug, called Rezulin, was touted as a promising treatment for people with adult-onset (or Type II) diabetes. Diabetes is a disease characterized by the body's inability to produce sufficient insulin, which is necessary for the breakdown of sugar. Insufficient insulin results in high levels of blood sugar, which if untreated can lead to heart disease and blindness. In the mid-1990s, several treatment regimes were available for Type II diabetes.

Rezulin was not effective, however, for Type I diabetes, where the body's inability to produce sufficient insulin is caused by a genetic abnormality. By stimulating the body's fat and muscle cells so that they absorb more sugar, Rezulin purportedly helps the body use its insulin more efficiently.

Warner-Lambert's submission of Rezulin as an NDA came at a time when the FDA was given authority by Congress to accelerate the review process for selected drugs. This policy is called the "fast track for drug review," and it was prompted in part by AIDS activists who lobbied to get potentially life-saving drugs to dying patients who had no other treatment options. The fast-track review was seen as a boost to drug manufacturers and as a source of hope for some victims of dread disease.

Adult-onset diabetes affects about fifteen million Americans, approximately 6 percent of the population. Another twenty-one million Americans are at risk of developing the disease. At the time of Rezulin's submission as an NDA, the market for a drug like it was over a $1 billion dollars a year. In 1994, the National Institutes of Health (NIH) launched a nationwide diabetes study involving more than four thousand volunteers in twenty-seven research sites across the country. Companies competed to get their drugs in the study. A drug that was demonstrably successful in this national study was assured a good position in the market. Warner-Lambert pledged to contribute

$20.3 million in exchange for the exclusive rights to any products that emerged favorably from the study.

Does this collaboration sound like an unusual one? A drug company pledges to contribute money to defray the costs of a government study that includes one of its drugs in exchange for the intellectual property rights of the discoveries made. Some might question whether this is too cozy a relationship between a drug company and a federal research center. How does a federal research center accept funding from a company that has financial stakes in the outcome of a trial without giving the appearance it is complicit with influence peddling?

On June 11, 1996, Warner-Lambert publicized that its new drug Rezulin was selected to be part of the largest diabetes study in the United States. The company news release quoted Dr. Richard C. Eastman, a leading diabetes researcher at NIH, as saying that Rezulin "corrects the underlying cause of diabetes." Eastman was director of the division of diabetes, endocrinology, and metabolic diseases at the National Institute of Diabetes and Digestive and Kidney Diseases (one of the NIH institutes), a position earning him an annual salary of $144,000 a year in the late 1990s (it also made him among the highest paid officials in the federal government).

Eastman was among a group of seasoned endocrinologists who expressed enthusiasm for Rezulin, supported its use in clinical trials, and also backed its adoption by the FDA in an accelerated review. In January 1997, Warner-Lambert's NDA for Rezulin was given final approval by the FDA, the first time the agency granted a fast-track review for a diabetes pill. In August of that year, the FDA approved a broader use of Rezulin, both as a stand-alone drug and in combination with other Type II diabetes drugs.

During the fall of 1997, several months after Rezulin had been on the market, the FDA began receiving reports from general medical practices that diabetes patients on Rezulin had experienced liver failure. It was estimated that more than a million people afflicted with Type II diabetes took Rezulin from the time it was introduced in March 1997 until June 1998, when NIH withdrew the drug from its national study. Britain announced Rezulin's withdrawal from the U.K. market in December 1997. The FDA, however, continued to support the drug with additional monitoring requirements until March 21, 2000, when the drug's registration was cancelled—twenty-nine months after the agency received the first reports of liver failure. By the time the FDA cancelled Rezulin, it had been linked to at least ninety liver failures, including sixty-three confirmed deaths and nonfatal organ transplants. Among those deaths was that of Audrey LaRue Jones, a fifty-five-year-old high school teacher from St. Louis who volunteered for the NIH study even though she did not have diabetes. After taking Rezulin for about seven months, Ms. Jones developed liver failure that required an organ transplant. She died after the transplant in May 1997. The NIH concluded that Jones' liver failure was probably caused by Rezulin.

In late 1998, the *Los Angeles Times* (*LA Times*) published a series of investigative reports that disclosed conflicts of interest among leading diabetes experts and drug companies, most prominently Warner-Lambert. Richard Eastman, while an employee of NIH, became a paid consultant to Warner-Lambert beginning in November 1995. According to reports in the *LA Times*, Eastman accepted at least $78,455 in compensation from the company. From 1991, Eastman then received at least $260,000 in consultant-related fees from a variety of outside sources, including six drug manufacturers.[18] Eastman acknowledged that he participated in a number of deliberations concerning Rezulin during the period he was both a consultant to Warner-Lambert and a federal employee. Among those deliberations were the selection of Rezulin for the national diabetes study (1994–1995) in addition to the decisions in 1997 on whether the drug should be retained in the study after patient deaths were reported.

Eastman was listed in company records as one of eleven members of the "faculty" of the Rezulin National Speaker's Bureau, a group of medical consultants paid for by Warner-Lambert who give talks to physicians on prescribing Rezulin to their patients. Eastman delivered the group's keynote address at its Dallas meeting on September 7, 1997. He also served for two years, including service on its editorial advisory board, for the National Diabetes Education Initiative, an organization financed in part by Warner-Lambert.

Senior employees of the Department of Health and Human Services (DHHS) are periodically monitored for potential conflicts of interest. Federal law makes it a crime for a federal employee to participate "personally and substantially" in governmental matters that affect an outside employer. The NIH ethics guidelines prohibits officials from consulting with private companies if it would "interfere in any way" with the official's public responsibilities. Neither federal law nor the NIH rules ban a government scientist from consulting, but the rules have considerable latitude for interpretation. Some observers found Eastman's multiple roles problematic—an opinion not shared, however, by his superiors.

LA Times investigative journalist David Willman learned that in June 1996, a lawyer at DHHS who reviews financial statements of senior federal officials expressed concern about Eastman's financial compensation from Warner-Lambert while he was directing a major study that included a product manufactured by a subsidiary of the company, Parke-Davis. The lawyer wrote Eastman and asked that he recuse himself from all official matters involving Parke-Davis (and therefore Warner-Lambert). Two of Eastman's superiors, one of whom was the director of NIH's Diabetes Institute, had already approved Eastman's consulting arrangement with Warner-Lambert after consultation with an NIH lawyer. Their decision indicated that Eastman has a right to engage in consultant work outside the institute in areas that relate to his government responsibilities.

Eastman was one of several scientists involved in decisions regarding the fate of a new drug that, at the very least, exhibited the appearance of conflict of interest.[19] Dr. Jerrold M. Olefsky was a prominent diabetes researcher who held a leadership position in the NIH's diabetes prevention program since 1994. Listed on three separate patents for Rezulin, Olefsky was the cofounder and president of a privately held firm that received substantial funding from Warner-Lambert. Olefsky was a strong proponent of including Rezulin in the national study, and he ultimately introduced Eastman to the compound and its promising preliminary results. In 1994, Olefsky was chairman of the NIH panel that reviewed the drugs adopted for the clinical trial. By mid-1995, the panel voted unanimously to include Rezulin. Later, in the summer of 1995, Olefsky was replaced as chairman of the panel because of concerns over the "appearance" of his potential conflict of interest in Rezulin. According to the *LA Times,* however, he was allowed to remain a member of the panel and serve on the study's steering committee despite that "appearance."

The *LA Times* probe found that at least twelve of the twenty-two scientists who played a central role in the NIH diabetes study received research funding or compensation from Warner-Lambert.[20] In addition, the *LA Times* uncovered a network of corporate influence peddling to scores of physicians and researchers: "Warner-Lambert and its affiliates paid speaking or other fees to more than 300 doctors, from endocrinologists to family practitioners. The company flew diabetes specialists to the 1996 Olympic Games in Atlanta and provided accommodations to the Chateau Elan Winery and Resort."[21]

The conflicts of interest involved in the introduction and adoption of Rezulin were brought to the attention of the public and Congress because the drug's failure aroused suspicion among journalists over, first, how and why the drug was adopted on the fast track by FDA and, second, why a number of deaths associated with liver failure did not prompt its removal from the market. Had the drug been successful, it is doubtful there would have been enough interest in the story for the media to have put resources into an investigative report.

As of this writing, no evidence exists to suggest that policy changes grew out of the disclosures associated with the Rezulin case. Like many conflict-of-interest cases, the Eastman–Rezulin connections were framed as idiosyncratic and atypical, which they almost assuredly are not as evidenced in the next case, which examines scientific advisors to government panels.

CASE 4: DANGEROUS VACCINE APPROVED IN PROCESS REPLETE WITH CONFLICTS OF INTEREST

The U.S. Food and Drug Administration (FDA) is considered by many to have the toughest standards among national health ministries for evaluating and licensing new drugs. As an example, the FDA took over a decade to review the efficacy and safety of the French abortion pill RU 486 for use in the United States after it had been approved in France and adopted by other European countries. Also, when the FDA refused to approve a drug called thalidomide, which was prescribed in the United Kingdom during the 1950s and early 1960s as a tranquilizer and a treatment for nausea during pregnancy, many American babies were saved from severe limb malformations.

The drug approval process involves many stages and multiple levels of review, both by FDA scientists as well as outside consulting scientists. Vaccines are a special category of drugs that are used for preventing rather than for treating an illness. The number of people who could potentially be vaccinated ranges from the entire population (such as during the polio epidemic) to a targeted at-risk subset of the population amounting to several million (as in a flu epidemic). When a vaccine causes harm, it does so against the prospect of preventing a disease. Most people who are vaccinated, let us say against the flu, ordinarily would not have been inflicted with the disease. From an ethical standpoint, the burden is high on the public health authorities and the vaccine producers to first and foremost "do no harm" to healthy people. For individuals who have been stricken by disease, the risk of drug use is always balanced against the benefits they may offer and the risks of doing nothing. When the possibility of disease is slight or its severity mild, assurances must be high that a vaccine is safe since the risks of doing nothing can often be very low.

A class of viruses called "rotaviruses" is one of the leading causes of acute gastroenteritis, a condition that can lead to severe diarrhea and account for about half of the cases among infants and children who require hospitalization. Approximately three million cases of rotavirus occur each year in the United States, and annual deaths resulting from complications of rotavirus range from twenty to one hundred.

Wyeth Lederle Vaccines and Pediatrics, a subsidiary of American Home Products, was the first pharmaceutical company to receive FDA approval for a rotavirus vaccine. The company filed its "investigational new drug application" in August 1987 for "Rotashield" vaccine, and it received approval in August 1998. Within about a year after the vaccine had been licensed, it was removed from the market after more than one hundred cases of severe bowel obstruction were reported in children who had received the vaccine.

When the circumstances behind the vaccine approval were investigated by a U.S. House Committee on Government Reform, it was learned that the advisory committees of the FDA and the Centers for Disease Control were filled with members who had ties to the vaccine manufacturers. In addition, it was also learned that conflicts of interest were endemic to the vaccine programs (see chapter 6). It is, of course, after a drug is withdrawn that we begin to understand the role that conflicts of interest are perceived to play in the complex process of weighing the benefits of a vaccine to a large population against the estimated risks of side effects to a relatively small population. When scientists who have personal financial interests in the vaccine are the key decision makers, the process is irreparably compromised.

These cases represent the scandals of their day. By themselves, they do not reveal the conditions that reinforce or replicate the unethical behaviors. How can we protect our revered institutions of higher learning, the institutions where we entrust the minds and shape the values of our future generations, from becoming knowledge brokers that serve the interests of their corporate sponsors, all the while alleging independence and objectivity? Every sector of our society, including business, government, professional organizations, the medical community, journals, and the universities have played a role in turning a blind eye to conflicts of interest. We begin our inquiry by examining the changes in federal policies that have transformed our universities into incubators for generating wealth and intellectual property while significantly compromising their virtue and public interest roles.

NOTES

1. This case was reported to me by Ingar Palmlund, who translated an article that appeared in the Swedish newspaper *Torsdag* ("Foretag styr allt fler professorer"), August 1, 2002.
2. Andrzej Górski, "Conflicts of Interest and Its Significance in Science and Medicine," *Science and Engineering Ethics* 7 (2001): 307–312.
3. Giovanni A. Fava, "Conflict of Interest and Special Interest Groups. The Making of a Counter Culture," *Psychotherapy and Psychosomatics* 70 (2001):1–5.
4. Arnold S. Relman, "The New Medical-Industrial Complex," *New England Journal of Medicine* 303 (October 23, 1980): 967.
5. Anonymous [editorial], "Medicine's Rude Awakening to the Commercial World," *The Lancet* 355 (March 11, 2000): 857.
6. Peter Gosselin, "Flawed Study Helps Doctors Profit on Drug," *Boston Globe,* October 19, 1988, A1.

7. Gosselin, "Flawed Study Helps Doctors," A4.
8. National Science Foundation, *Science and Engineering Indicators 2002: Academic Research and Development,* ch. 5, table 5.2. Support for Academic R&D, by Sector: 1953–2000.
9. *Drug Intelligence and Clinical Pharmacy* 20 (1986): 77–78.
10. Drummond Rennie, "Thyroid Storm," *Journal of the American Medical Association* 277 (April 16, 1997): 1238–1243, at www.ama-assn.org/sci-pubs/journals/archive/jama/vol_277/no_15/ed7011x.htm.
11. Rennie, "Thyroid Storm."
12. Karen Kerr, "Drug Company Relents, Dong's Findings in *JAMA*," *Synapse* (May 1997): 1, at www.ucsf.edu/~synapse/archives.
13. Kerr, "Drug Company Relents," 1.
14. Kerr, "Drug Company Relents," 1.
15. *State of Tennessee v. Knoll Pharmaceutical Co.*
16. G. H. Mayor, T. Orlando, and N. M. Kurtz, "Limitations of Levothyroxine Bioequivalence Evaluation: An Analysis of an Attempted Study," *American Journal of Therapeutics* 2 (1995): 417–432.
17. Ralph T. King, Jr., "Judge Blocks Proposed Synthroid Pact, Criticizing the Level of Attorney's Fees," *Wall Street Journal,* September 2, 1998, B2.
18. David Willman, "Waxman Queries NIH on Researcher's Ties," *Los Angeles Times,* December 9, 1998.
19. David Willman, "2nd NIH Researcher Becomes a Focus of Conflict Probe," *Los Angeles Times,* September 4, 1999.
20. David Willman, "Scientists Who Judged Pill Safety Received Fees," *Los Angeles Times,* October 29, 1999.
21. David Willman, "The Rise and Fall of the Killer Drug Rezulin," *Los Angeles Times,* June 4, 2000.

3

UNIVERSITY–INDUSTRY COLLABORATIONS

Over the past several decades, the goals, values, and practices of American research universities have become transformed in ways that have brought them in greater alignment with industrial interests. These changes are comparable in their transformative impacts to the changes that universities experienced during two other periods in the twentieth century: postwar science in the 1950s and the Vietnam War years from the mid-1960s to the early 1970s.

According to Geiger's history of the American research universities prior to the 1940s, universities and industry carried out their research on parallel tracks, interacting in only minor ways.[1] In special cases, industry supported a few university-based institutes in applied research during the 1920s. University researchers did some consulting; otherwise, corporations made no substantial investments in university research.

After World War II, federal funding for science began to rise steeply. The trend of federal sponsorship for research resulted in the modern research university. In 1940, the total funds available for scientific research in universities from all sources was $31 million. Forty years later, the figure has surpassed $3 billion—a hundredfold increase. "Whereas before World War II university research had been largely . . . a privately-funded research system, public support came to dwarf foundation and other private support after the war."[2] Up until the early 1960s, universities received between 6 and 8 percent of their basic and applied research budgets from industry. The figures dropped precipitously during the mid-1960s until the early 1970s, when university research support from industry was in the 2 percent range, the lowest it had been in twenty years. Federal support for university research concurrently rose during this period, making it less attractive to corporate funding; meanwhile, industry invested in its own research capacity. As the rate of increase of federal research support declined during the late 1970s and early 1980s, universities were once again looking for corporate research dollars.

Other changes that took place in research universities were brought about by the U.S. involvement in the Vietnam War, a time when many institutions debated the role of the military influences on higher education, including reserve officer training programs (ROTC), weapons development, and classified research. Some of the nation's leading academic institutions reexamined their roles in society and rejected classified research and contracts involving weapon systems. "In the universities, the potentially conflicting demands for purity and relevance were reconciled by focusing on poverty, race, urban problems, and environmental protection. No proponent of social responsibility yet argued that conducting research for industry would serve a useful social purpose, that the development of new products or the improvement of manufacturing techniques had a bearing on the public weal."[3] What continued to exist, however, were research relationships and consultantships between industry and departments of chemistry as well as schools of agriculture and engineering.

In 1968, James Ridgeway published *The Closed Corporation,* a book that exposed the myth behind the "ivory tower" of academia. He compiled a rich tapestry of cases showing how professors set up their own companies and use public resources for private gain. Ridgeway described how faculty at elite institutions, in disciplines such as economics and business, established new kinds of knowledge companies based on techniques of systems analysis, statistics, mathematical modeling, and behavioral psychology to sell "social problem solving," a concept that brought handsome fees from corporations. The new professor–entrepreneurs were beginning to change the character of the modern university. "As power brokers, the professors act with one hand in the university and the other in a big corporation; they move in and out, using their prestige as scholars to advance the interests of the company; or on the other hand, using their influence with the company to help the university get research funds."[4] In the next twenty years, the changes in the commercial role of professors and their universities would dwarf what Ridgeway discovered had occurred between the 1950s and 1960s.

IVORY TOWER OBSTACLES TO GLOBAL COMPETITIVENESS

By the beginning of the 1980s, a more fiscally conservative and antiregulatory influence had taken hold of Congress and the presidency, and with it came a changing attitude toward universities. The new centers of political power had a love–hate relationship with American higher education. Universities were seen both as the cause and the savior of America's declining economic competitiveness.

On the negative side, universities were being attacked on several fronts. Fiscal conservatives blamed universities for charging government too much for overhead on sponsored grants and for misappropriating government funds. Some scientists were mocked for their irrelevant and wasteful studies during the thirteen years (ending in 1988) that Senator William Proxmire bestowed his "Golden Fleece Award" for the federal research grant that he ranked highest on the abuse of taxpayer money. At the same time, a new group of conservative ideologues attacked universities for their "liberal agendas," which included social engineering and artistic expressions that failed to conform to "traditional" family values.

On the positive side, universities were viewed as an underutilized resource. Too many useful ideas and innovations fester in academic journals and never become transferred into America's vast industrial system, where it can improve efficiency, provide consumer products, and create new wealth. The United States was perceived as losing its economic leadership among the nations of the industrial world, and the reason given was that Americans were either slow or negligent in turning basic scientific discoveries into applied technology. Two reasons were cited for why the United States was not innovating fast enough. First, too many regulatory obstacles were in place—a response to the rising concerns over the environment and public health. Second, too great a lacuna existed between universities and industry. This separation meant that potentially valuable discoveries would never be developed.

Certainly, a case could be made about America's industrial decline. We were losing out in automobiles, steel production, microelectronics, and we were experiencing strong competition in other areas, such as computer technology and robotics. According to a report prepared for the Association of American Universities, "Between 1966 and 1976 the U.S. patent balance declined with respect to the U.K., Canada, West Germany, Japan and the USSR. By 1975 it was negative for the last three. The proportion of the world's major technological innovations produced by the United States decreased from 80 percent in 1956–58 to 59 percent in 1971–73."[5] It soon became a fashionable idea among policy makers that the U.S. competitive position could be improved by fostering greater collaboration between academia and industry.

Rosenzweig's account in *The Research Universities and Their Patrons* states: ". . . there is good reason to believe that productivity is linked to scientific and technological innovations, that universities are a leading source of these valuable products, and that closer connections between universities and industry may help to move scientific research into practice."[6] Capturing the mood at the time, Rosenzweig notes: "It is clear that many thoughtful people are beginning to place their hope for improvements in the competitive position of American business and in the health of university-based science in the growing collaboration between business and universities."[7] Efforts were made at the National Science Foundation in 1972 to build stronger industry–university partnerships, but they were considered by congressional leadership to be unstructured, unfocused, and uncoordinated. Despite such initial attempts, the idea of building bridges between universities and industry was gaining advocates across the political spectrum.

President Reagan's science adviser, George Keyworth II, was a strong proponent for university–industry partnerships as a salvo for improving America's declining competitive position. For Keyworth, one of the failings of American science was that "most academic and federal scientists still operate in virtual isolation from the expertise of industry and from the experience, and guidance of the marketplace."[8]

During the 1980s, a series of federal and state policies established incentives for private companies to invest more heavily in university research, a move that provided opportunities for universities to benefit directly from the discoveries of their faculty. The two basic approaches—namely, university–industry partnerships and patenting—are encapsulated by the phrases "technology transfer" and "intellectual property rights of basic research."

A decade of aggressive university–industry partnerships was stimulated, in part, by the Supreme Court in *Diamond v. Chakrabarty* (1980), which ruled that genetically modified bacteria were patentable in-and-of-themselves, apart from the process in which they are used. This ruling opened up the floodgates for the patenting of cell lines, DNA, genes, animals, and any other living organism that has been sufficiently modified by humans to qualify as "products of manufacture." With this ruling by the Supreme Court, the U.S. Patent and Trademark Office extended intellectual property rights to segments of DNA whose role in the organism was not understood. This decision meant that university scientists who sequenced genes had intellectual property that they could license to a company or that could serve as the catalyst for forming their own company.

Also in 1980, the National Science Board made university–industry partnerships a focus of study. Congress amended the patent law in the Patent and Trademark Amendments Act of 1980, more commonly known as the Bayh–Dole Act (PL 96-517), giving universities, small business, and nonprofit institutions title to inventions made with federal research funds regardless of which agency funds had been used to make the invention. The eligibility for obtaining title to inventions that were derived from federally supported work was extended to industry as a whole by Executive Order (12591) on April 10, 1987.

While the Bayh–Dole Act was the most visible of the new federal policies, a host of other laws and executive orders reinforced the philosophy behind the act. The Stevenson–Wydler Technology Innovation Act of 1980 (PL 96-480) was designed to foster technological innovation in the United States by encouraging cooperation between industry, government, and universities. In 1981, the Economic Recovery Tax Act (PL 97-34) gave tax credits to companies for their contributions of research equipment to universities. The act also allowed Research and Development Limited Partnerships (RDLPs) to be eligible for favorable tax treatment if they were designed with university–industry collaborations in mind.

As a result of these acts, the university–industry relationship took a major leap in its evolution. In 1984, Erich Bloch, the first director of the National Science Foundation appointed from private industry, set up the National Academy of Engineering and established research centers at universities to foster cooperation between academia and industry. In addition, the emergence of state economic development programs, many in the field of biotechnology, provided incentives for university–industry cooperation. During the 1970s and 1980s, the formation of University–Industry Research Centers (UIRCs) was supported by funding primarily from federal and state governments. Before 1980, only three states had UIRCs. Ten years later, UIRCs were operating in twenty-six states. By 1990, research at these university–industry collaborative centers comprised about 15 percent of the total research and development budget of universities.[9]

According to a study by the now defunct congressional Office of Technology Assessment (OTA), "The confluence of events and policies increased the interest of universities, industry and government in activities pertaining to partnerships between academia and business in all fields of science."[10] The OTA study anticipated some of the intractable problems that would result from this aggressive effort to merge industry and university interests. "It is possible that the university–industry relationships could adversely affect the academic environment of universities by inhibiting free exchange of scientific information, undermining interdepartmental cooperation, creating conflict among peers, or delaying or impeding publication of research results. Furthermore, directed funding could indirectly affect the type of basic research done in universities, decreasing university scientists' interests in basic studies with no potential commercial payoff."[11] The prediction by the OTA proved to be prophetic.

Throughout the decade, the growth of university–industry relationships had been unprecedented—a fact specifically evident in biotechnology, where company-sponsored university-based research was 20 percent higher than the overall average for all industrial sectors, with nearly 50 percent of biotechnology firms supporting research in universities.[12] During that period, at least eleven multiyear, multimillion-dollar contracts for research in biotechnology were issued by chemical and pharmaceutical firms. By 1984, industry support for biotechnology in universities totaled $120 million, which was estimated to be about 42 percent of all industry-supported university research.

The impact of the Bayh–Dole Act on university patenting was highlighted in the summary statement of a workshop on intellectual property rights held by the National Research Council:

> University patenting steadily increased from 1965 to about 1980, when there was a sharp increase in patenting that has continued into the 1990s. From 1965 to 1992, university patents increased by a factor of over 15 [1500 percent], from 96 to 1500, whereas total patents increased by only about 50 percent. By the year 2000, universities had been awarded over 3,200 patents. The greatest portion of the increase in university patenting has been in the biomedical sciences, and many university patents cover inventions that are useful primarily for scientific research.[13]

Certain members of Congress began to express reservations about the merging of industry and university research agendas in addition to the promotion of new collaborative arrangements. First was the issue of the proper use of public finds.

With the commingling of university and industry-supported research, Congress questioned whether public funds inadvertently would be used to advance industry agendas. Will the new partnerships lead to misconduct and conflicts of interest? Were taxpayers getting a good return on their investment by giving away to the private sector the intellectual property derived from federally funded research? According to Slaughter and Leslie, "The United States is the only country in which universities hold title to intellectual property developed by faculty with federal grants."[14]

A series of congressional hearings took place between 1981 and 1990 to investigate these and other questions. One of the earliest hearings on university–industry arrangements was cochaired by the young congressman from Tennessee, Al Gore. In his opening remarks, Gore outlined the growing problems with the policies enacted to accelerate commercialization of biomedical research: "We return with a continuing concern that our universities, the source and foundation of these technologies, may be permanently altered by the increasing number of commercial agreements they are developing."[15] Gore cited the issues facing the universities that became partners with industry: conflicts of interest, allocation of resources, teaching graduate students, shifting research priorities, publication of research results, and the struggle to remain economically solvent. Doug Walgren, the other cochair, asked the central question of the hearings: "Can universities successfully preserve the free exchange of ideas between students and faculty while meeting the obligations of industrial arrangements?"[16]

Eight years later in 1990, Representative Ted Weiss chaired subcommittee hearings that exposed a litany of cases illustrating the downside of university–industry collaborations. Weiss cited research in social psychology, suggesting that where large gifts may influence behavior, small gifts are likely to influence attitudes. The pervasive use by drug companies of small gifts and travel grants to academic physicians is designed to shape a positive attitude toward the company and its products. But the Weiss subcommittee issued its strongest recommendations against conflicts of interest in drug evaluations and requested the Department of Health and Human Services (DHHS) to "immediately promulgate PHS [Public Health Service] regulations that clearly restrict financial ties for researchers who conduct evaluations of a product or treatment, in which they have a vested interest."[17]

The next decade brought a bounty of cases (too numerous to document) of academic scientists following incentives established by federal and state governments. These scientists were aided by either the willing participation or the passive tolerance of university administrations, and they ultimately became involved in ethically questionable entrepreneurial ventures. Investigative journalists from local and regional newspapers brought many of these cases to national attention. I have selected some of the more demonstrative stories to illustrate the range of the new partnerships between universities (or government) and industry.

PUBLIC UNIVERSITY DEVELOPS AND MANUFACTURES A DRUG

Imagine a situation where medical researchers at a public university discover a drug, develop it for their patients, manufacture and sell it illegally to other medical practices, and build a publicly financed facility to keep up with the demand. If it seems a little farfetched, we have only to go to the University of Minnesota to understand how such a scenario came about.

In 1970, the Food and Drug Administration granted Investigative New Drug (IND) status to the University of Minnesota (UM) for the experimental drug Antilymphocyte Globulin (ALG). Made from horse tissue, the product was used in organ transplants to prevent the new host's rejection of the foreign organ. How it works is that human cells are injected into horses; afterward, the animals are bled and their serum is extracted. The drug showed considerable promise in suppressing the body's immune system to therefore reduce the risks of organ rejection. ALG was developed by members of the Department of Surgery of the University of Minnesota Medical School, and it was used for two decades. However, the drug carried only experimental status and was never approved by the FDA for general use. According to John Najarian, the head of the surgery department at UM during that period, ALG was manufactured by the university for twenty-two years and was applied in the treatment of more than one hundred thousand transplant patients in over one hundred medical centers and hospitals.[18]

Two journalists from the *Star Tribune,* Joe Rigert and Maura Lerner, followed the case closely in the early 1990s. They wrote: "In the ALG case, the university operated the equivalent of a small drug company of its own. It even built a new $12.5 million facility for the program to be funded from the sale of ALG."[19]

Faculty of the UM Medical School set up their own private nonprofit corporations, which brought in more than $4 million since the mid-1980s. One of these, Biomedical Research and Development (BRAD), was doing contract work, collecting fees, and distributing the funds without the traditional university oversight. The investigative journalists found glaring conflicts of interest in this operation. One university scientist at BRAD was a consultant to a company that was sponsoring tests in his laboratory. A second entity at the University of Minnesota, which operated much the same way, was the Institute for Applied and Basic Research in Surgery (IBARS). This faculty-based entity raised money from corporations to conduct research.

In 1992, FDA inspectors on a site visit to the university filed twenty-nine violations on the ALG program. These violations included the following: failure to report adverse reactions; failure to monitor studies; unauthorized exports of the drug; numerous gaps in the testing record; and improper claims that the drug's safety had been established.[20] By August 1992, the FDA took action to stop the university's sale of the multimillion dollar transplant drug. The surgery chairman of UM sent letters to transplant centers warning them not to start new transplant patients on the drug because of concerns about side effects. He also announced the recall of one lot of the drug because of possible bacterial contamination resulting from leaky vials. The director of the ALG program was fired after it was learned he had personally pocketed tens of thousands of dollars from products he sold to a Canadian company. The director had the title "associate professor of surgery," but, interestingly, he did not hold a Ph.D. or a medical degree.

The drug law is explicit. No biologic product may be shipped across state lines for sale, barter, or exchange unless it is manufactured and prepared in an FDA-licensed facility. The surgery department at UM set itself up as the developer, evaluator, manufacturer, and distributor of a drug used to save lives. It did so by circumventing university procedures and violating federal laws and regulations.

The development, marketing, and sale of drugs are not part of the university's charter. It draws the university into a world where cutting corners for the bottom line, taking legal shortcuts, and standing in opposition to regulation are commonplace. One unsavory step by an entrepreneurial faculty in this environment of academics and business can cast a broad dark shadow over the entire institution.

MULTINATIONAL CONTRACTS WITH UNIVERSITY OF CALIFORNIA, BERKELEY

During a talk I delivered to the Environmental Grantmakers Association in the fall of 2000, I fancifully proposed the following futuristic newspaper headline: "Monsanto Purchases Princeton University; Promises to Maintain its Nonprofit Status." For most people, the sell-off of a major private university to a for-profit corporation seemed too farfetched for serious consideration. The thought of it conjures up a cultural taboo likened to the privatization of Yellowstone National Park. Others to whom I have spoken (who have followed the trends in university entrepreneurship) do not view the prospects as so improbable. The University of California at Berkeley (UCB) has elevated the issue of university–industry alliances to a new level of discomfort.

Gordon Rausser became dean of the College of Natural Resources at UCB in 1994. His appointment came at a time when the private sector's support of the University of California system was growing 15 percent annually while the public sector's support was growing at a mere 6 percent annually. As a result of the state's budget cuts, his college received only 34 percent of its budget from state funds. Rausser, an economist, therefore came up with a plan to get an infusion of corporate money that ostensibly auctioned off the intellectual resources and reputation of Berkeley to the highest bidder. He sent out letters of inquiry to sixteen agricultural, biotechnology, and life-sciences companies indicating that his college would accept formal proposals for a research partnership. He received positive responses from Novartis, DuPont, Monsanto, Pioneer Hi-Bred, Sumitomo, and Zeneca. Formal proposals were received from Monsanto, Novartis, and from DuPont and Pioneer jointly. The college selected as its research partner Novartis, a $20 billion food and pharmaceutical company that was building its future on biotechnology.

The result was a five-year, $25-million alliance between Novartis Agricultural Discovery Institute (NADI), a subsidiary of Novartis, and the Regents of the University of California on behalf of the Department of Plant and Microbial Biology in the College of Natural Resources. In this far-reaching and unprecedented agreement, all members of the department were given an opportunity to sign on. By December 1998, thirty out of thirty-two faculty members either signed or were expected to sign on to the agreement. The *Chronicle of Higher Education* noted that "the deal is unusual because it applies to an entire department, not a single researcher or a team working on a specific topic."[21]

Two-thirds of the funds ($3.33 million per year) were to be used for unrestricted research, which the administrators showcased to emphasize the protection of academic freedom. One-third of the funds ($1.67 million per year) would be used for overhead and infrastructure costs, which the dean emphasized was a clear win for a university that was struggling to compensate for declining public support.

The agreement established two oversight committees. One five-member committee, which consisted of two individuals from the NADI, would decide which projects get funded. So while the private institute would not define the projects, it would have strong input into (and exert direct influence on) which research would get funded. The second committee, a six-member advisory committee responsible for managing the relationship between the university and NADI, comprised three members associated with Novartis and three from the university, two of whom were associated with the agreement and a third who was a UCB faculty member outside the Department of Plant and Molecular Biology.

What did Novartis get for its $25 million? It bought the "first right to negotiate a license" from any discoveries derived from Novartis research funds granted to the university or from collaborative projects between the institute and UCB scientists. Under the agreement, the university would own all patent rights if an invention were developed as a joint effort by NADI and a UCB employee. Alternatively, if an invention were made by a NADI employee using university facilities, it would be jointly owned.

Company researchers sit on the UCB internal research committees. This portal to the university gives the company a shaping function over the research it funds. For example, the likelihood that any member of the Department of Plant and Molecular Biology would investigate the negative impacts of a Novartis product is remote.

The agreement set restrictions on all members of the department who were cosigners. NADI would ultimately be financing 20 percent of the department's budget. The participating faculty would then have the opportunity to sign a confidentiality agreement that would give them access to NADI's proprietary genetic database. Once a faculty member signed the confidentiality agreement, he or she could not publish results that involved the data without approval from Novartis.

In May 2000, the California Senate, led by Senator Tom Hayden, held hearings on the Berkeley–Novartis contract.[22] Dean Rausser was asked the following: Suppose a professor who has signed a confidentiality agreement comes across data that represents a serious danger to the public and wishes to speak out as a matter of conscience. Will UCB come to the scientist's aid? The answer was unambiguous. The university had no obligation to defend scientists who break the contract, even if there is a public interest in revealing information.

Under the agreement, NADI could request publication delays of up to 120 days to ensure that it could develop patent applications for new discoveries. Novartis had the right to ask the university to patent certain inventions at the company's expense. Once the patent was secure, Novartis had the right to negotitate an exclusive license to develop a product from the patented material. The agreement also called for restrictions on the exchange of certain biological materials. Any scientific exchanges of information outside the inner circle of NADI partners would have to be approved by Novartis. The sad result of this precedent-setting agreement is that other universities are now signing similar agreements that involve creating extensive roadblocks against the free flow of scientific information to protect the interests of the corporate partner.

At the California Senate hearings, Senator Hayden said that the UCB agreement may compromise the unique role of a public university to undertake independent research on the negative effects of genetically engineered crops. He called upon the legislative body to enact a law that would restrict the conflicts of interest of university faculty who are involved with the biotechnology industry. Hayden's concerns were not without historical precedent. In 1969, state and federal agencies focused on emergency environmental problems arising from the massive oil leak into the Santa Barbara Channel from Union Oil Company's offshore well. At the time, the oil industry had a cozy relationship with university experts in geology, geophysics, and petroleum engineering. State officials could not get local experts to testify at hearings in the half-billion dollar damage suit against Union and three other oil companies. According to California's attorney general, the explanation they offered for the lack of cooperation was that "petroleum engineers at the University of California campuses of Santa Barbara and Berkeley and at the privately supported University of Southern California indicated that they did not wish to lose industry grants and consulting arrangements."[23]

Although California is a state that already has some of the most stringent disclosure rules for public university employees, the Senate hearings had no visible impact on the nature of the agreement or on California's policies on conflict of interest. What they did accomplish, however, was to reinforce the criticisms of some students and faculty of the perils of the university–industry interface.

The hearings also brought national attention to the growing problem of institutional conflicts of interest—that is, when a university has multimillion-dollar, long-term interests and equity growing out of a private alliance with a for-profit company. Some products of Novartis will eventually have the imprimitur of UCB. How long will it take before a genetically modified food crop is developed, tested, and patented at UCB and marketed globally by Novartis? One UCB biologist and critic of the agreement asked: "Will universities cease to serve as places where independent (environmentally and socially) critical thought is nurtured?"[24]

In the fall of 2001, UCB biologist Ignacio Chapela and his graduate student David Quist published a paper in *Nature* stating that the authors found native strains of Mexican corn contaminated with the DNA from bioengineered strains.[25] This finding was a particularly concerning result because genetically modified (GM) seeds were prohibited from being planted in Mexico, although GM corn from the United States was exported to Mexico for processing. It is possible that some of that corn was planted or that virgin GM seeds were grown in violation of the regulation. *Nature* received numerous critical comments from other scientists, some stating that Quist and Chapela's conclusion was nothing more than an artifact of the measurement technique—a process called "inverse polymerase chain reaction." *Nature* apologized to its readers for publishing the article but did not retract it. Instead, it invited comments. In a letter to *Nature,* six scientists from the UCB Department of Plant and Microbial Biology questioned the Quist-Chapela results.[26] It was also well known that Chapela was a vocal critic who argued against the Berkeley-Novartis contract. The special connection between some critics and the biotechnology industry raised suspicions that some of this criticism was payback for Chapela's opposition to Berkeley's partnership with a major producer of GM products.[27] Critics disputed the allegations and claimed that their criticism was just science at work, nothing more.

Two independent teams, commissioned by the Environment Ministry of Mexico, undertook their own studies of the Quist-Chapela findings that native corn harvested in Oaxaca, Mexico, had been contaminated by genetic segments unique to GM varieties. Mexico's National Institute of Ecology released a statement in August 2002 that confirmed the contamination.[28]

What needs to be shown is that the trend at privatizing the production of knowledge in the country's colleges and universities is affecting the fundamental values of academic research, its objectivity, and its soundness; otherwise, these changes appear to be coming at no cost to America's research community and to the broader public they serve. Perhaps these effects become more sharply in focus when America's elite universities compromise their independence as they serve unabashedly as legitimating voices for well-endowed corporations. A policy center situated at Harvard's School of Public Health is a case in point.

HARVARD'S CENTER FOR RISK ANALYSIS

In March of 2001, Public Citizen, the nation's premier consumer organization, issued a report that revealed how major polluting corporations gained access to and influence over a Harvard University policy center. The 130-page report titled *Safeguards at Risk: John Graham and Corporate America's Back Door to the Bush White House* examined the record of John Graham, founding director of Harvard's Center for Risk Analysis (HCRA) and the work of the center.[29] During the month that Public Citizen released its muckraking report, Graham was nominated by President George W. Bush to assume the head of the Office of Information and Regulatory Affairs of the powerful Office of Management and Budget (OMB).

Public Citizen laid out the corporate funding sources of the HCRA and argued that a direct connection existed between the corporate interests that funded the risk center and the studies that were produced by the center. For example, Graham solicited financial contributions from a cigarette company while also downplaying the risks of secondhand smoke. The center published a study funded by AT&T Wireless Communications opposing the ban on using cellular phones while driving. The center's newsletter (*Risk in Perspective*) downplayed the risks to children from exposures to pesticides and plasticizers (bisphenol-A and phthalates) without alerting the readers that the center is funded by the manufacturers of those chemicals.

According to Public Citizen, the HCRA gets it funding from more than one hundred large corporations and trade associations, including DOW, Monsanto, DuPont, as well as major trade groups such as the Chlorine Chemistry Council and the Chemical Manufacturers Association (renamed the American Chemistry Council). Most of these corporations were reported to have made unrestricted grants to the center while others such as the American Crop Protection Association and the Chlorine Council provided restricted grants.

Whether or not corporate entities give restricted or unrestricted grants to universities is not the question; the issue is that gifts, grants, and contracts do not come to universities without some benefit to the corporate donors. In the case of Harvard's risk center, there was a major campaign to bring in significant sums of corporate money. These hundred or so corporations were surely not interested in pure research. Faculty and administrator entrepreneurs, like market analysts, identify what corporations want and then shape the center's goals to match those interests. Alternatively, a center with an ideological focus of special appeal to corporations will find its solicitations of support received more favorably. For example, the HCRA promoted ideas about regulation that were consonant with many sectors of the corporate community, such as use of the market more forcefully in lieu of regulation, quantitative risk analysis (which sets a high burden of proof for regulatory action), comparative risk analysis (which conflates voluntary and involuntary risks), and cost-benefit analysis (which ranks human health risks against corporate profits).

In a letter to EPA administrator Carol M. Browner dated April 19, 2000, Consumers Union criticized the HCRA study on the economic effects of pesticide regulation that would result from the implementation of the Food Quality Protection Act (FQPA) of 1996. The Harvard risk group claimed that implementation of the FQPA could result in up to one thousand premature deaths per year due to decreased food consumption resulting from higher costs. The FQPA was one of the least-contested environmental acts enacted in decades, and it ultimately became a victory for multistakeholder negotiations. Passed unanimously by both houses of Congress, the FQPA was supported by fifty-five interest groups, including representatives of the agricultural industry and environmental organizations.[30] Consumers Union noted that the HCRA study was paid for by the American Farm Bureau Federation, a trade organization that waged a major campaign against the FQPA.

John Graham solicited money from Philip Morris around the time he could offer the company a critique of the view held by the EPA that secondhand smoke was a health hazard. When Graham learned that the Harvard School of Public Health had a policy against accepting cigarette money, he returned the check to Philip Morris. Eventually, the HCRA did receive funding from Kraft, a subsidiary of Philip Morris.[31]

I have heard it said by some university administrators that it is better for universities to be funded by many corporations than by a few since that scenario will ensure that a narrow set of private interests do not shape the value of the university. Harvard's risk assessment center had scores of corporate sponsors, many of whom shared the same interests as the center in reducing regulations and in opposing the use of the "precautionary principle" on health and safety. The diversification and breadth of its industry support in this instance shows that the HCRA had positioned itself on issues of regulation, health, and safety that were ideologically biased toward the interests of its sponsors. As *Washington Post* columnist Dick Durban noted, "Environmental regulations mainly show up in Graham's research as a waste of money, and as such, the choice of regulations make us responsible for 'statistical murder' by drawing resources away from other, more cost-effective programs."[32]

Academic policy centers like the HCRA are more vulnerable to industry influence from funding sponsorship than science and medical departments. The reason is that policy analysis involves the use of discretionary assumptions and ideologically based frames of analysis that are often masked by quantitative symbols, technical jargon, and the selective use of supporting data that, in concert, are partial to certain outcomes. For example, if efficiency means saving money, then an extra cost of $1 million to reduce the pollution that inflicts a lung disease on eighty people with a lifetime treatment cost of $10,000 per person, is not justified.

Companies may continue to fund scientific research at universities even when the results are unfavorable to one of their products. Of course, certain cases exist where companies exerted pressure on scientists to get the results the companies wanted; if the scientists did not succeed, the company withdrew further support. But many continue to support academic science because they buy into the standards of the scientific method, which are required to be used under regulation. Canonical techniques in science are widely shared, and many experiments are subject to replication. The same is not true for policy research, where standardized methodologies are rare, if they exist at all, and data are often massaged to deliver results that pleases a partisan sponsor. Journalist Durbin wrote, "Graham's controversial research always seems to wind up with the same conclusion: We don't need more regulation in the private sector."[33]

Assume, for instance, that the American Farm Bureau Federation were interested in a critique of a new bill that would strengthen pesticide regulations. It would be the height of folly to believe that it would give money to the Harvard risk center without some assurance that it would receive in return results that conformed to its financial interests (or else such results would not see the light of day). In this respect, policy study centers, like law firms, are much more vulnerable to influence by corporate funding sources than are academic science departments. Some are less obvious than the HCRA about their ideological leanings. It worked out for John Graham, whose corporate entrepreneurship landed him the OMB position in May 2001, over the many objections that the center he directed at Harvard was a shill for private interests.

BLUNTING OF CRITICISM IN ACADEMIA

I was a young and somewhat politically naïve untenured assistant professor at Tufts University in the 1970s. Toward the end of the decade, I became acting director of a fledgling policy program. With a new grant from the Fund for the Improvement for Post Secondary Education, I led an innovative class in which students worked as a team to investigate a local toxic-waste controversy. The case involved the W. R. Grace Company and the town of Acton, Massachusetts, a suburb of Boston. Two wells were closed in Acton during December 1978 because of chemical contamination. The town had reason to believe the contamination came from production facilities at the Grace chemical plant. In exchange for the right to extend its manufacturing site, Grace provided funds for the community to contract a study that would investigate the source of pollution. On completion of an independent hydrogeological study of the toxic contamination, town officials had evidence that Grace was the source of some of the chemical contaminants. The task of the students was to do the following: perform a background analysis of Grace's environmental record; conduct an evaluation of the contested hydrogeological studies on the impacted land and acquifers; study the toxicological information about the chemicals discovered in the contaminated wells and aquifers; and investigate federal and state enforcement actions taken and the company's response. They did everything that well-trained young scholars would do. They conducted interviews; they reviewed government documents; and they collected primary data, carefully citing all their sources.

By the time the final report was in preparation, word had gotten to the W. R. Grace Company that a study would be released under my supervision and that it was not favorable to the company's public image. A vice president of the company made an appointment to meet with the president of Tufts University, who at the time was Dr. Jean Mayer, the internationally acclaimed nutritionist. Mayer wrote frequently and voluminously about the social responsibility of government to address hunger and to especially ensure adequate nourishment to pregnant mothers and newborns living under poverty. He was one of the architects of the Women, Infants, and Children (WIC) program, which provides food stamps for women and children.

The purpose of the visit by Grace's vice president was to stop the release of my students' report, which was titled "Chemical Contamination of Water: The Case of Acton, Massachusetts," and to cast disparagement on me as their advisor. Perhaps it was his way of telling the president that Tufts could do without professors like Krimsky, who are "unfriendly to corporations." Fortunately for me and my students, President Mayer was not persuaded by the visitor's remarks and reminded Grace's representative that the university must protect the academic freedom of faculty to educate, write, and speak as they deem appropriate for their discipline.

This situation was one in which Grace had no foothold at Tufts. The corporation was not represented on the Board of Trustees. It had not contributed any substantial grants to the university. Moreover, as I discovered some years later, Jean Mayer was on the Board of Directors of the Monsanto Corporation, a competitor to the W. R. Grace Company in chemical manufacturing.

It has occurred to me, on more than one occasion, that in conflicts between academic freedom and money, universities will usually forego the former when the financial stakes are high enough. Without guaranteed tenure, a faculty member is dispensible. The trade-off is easily rationalized by administrators who argue that money, not academic freedom, makes a great university.

No one really knows how often these trade-offs are made. All we have are anecdotal cases. University administrators are not likely to admit that they have dismissed faculty or refused to hire applicants for speaking out on issues that embarrass important funders.

For example, David Healy was offered a $250,000 a year job at the University of Toronto as clinical director of the Centre for Addiction and Mental Health and as professor of psychiatry. In November 2000, Healy gave a talk at the Centre, based on his and other's research, where he stated that Prozac could cause some patients to commit suicide. The manufacturer of Prozac, Eli Lilly, was a major donor to the University's teaching hospital. Representatives of Eli Lilly were said to have been at the conference.

Several weeks after his talk, Healy received a letter from the physician-in-chief at the Centre stating that the offer of his appointment was withdrawn because it was learned that he (Healy) was not a good fit as a leader of an academic program. Healy filed a multimillion dollar lawsuit against the University of Toronto and its teaching hospital for revoking his contract one week after he gave a lecture on the link between Prozac and suicide, ostensibly asking the Canadian courts to rule on his alleged denial of academic freedom.[34] The official statement of the Centre regarding its rescinding of Healy's job was that its hiring decisions have never been influenced by outside donors. By May 2002, Healy's suit resulted in a settlement for an undisclosed sum, and the university appointed him visiting professor of its faculty of medicine. The joint statement issued by the university and Healy maintained that although the psychiatrist believes his clinical appointment was revoked because of the speech on November 2000 criticizing Prozac, he "accepts assurances that pharmaceutical companies played no role" in the decision to withdraw his job offer.[35] In situations of this nature, the pharmaceutical company may indeed have played little or no role in deciding the fate of Healy. Its values and interests may already have been internalized by the administrative heads of the hospital and university, who needed little prodding to understand their sponsor's concerns.

Another scenario draws attention to the uneasy relationship that has always existed between hospitals and medical schools. Clinical faculty usually hold appointments in both institutions, each of which may function under different ethical rules. Efforts to harmonize those rules have not been without contention since hospitals and medical schools are independent entities with separate (albeit overlapping) Boards of Directors or Trustees, which cooperate for their mutual benefits.

David Kern is an occupational health physician who was hired at Memorial Hospital in Pawtucket, Rhode Island, where he was chief of general internal medicine and head of its Occupational and Environmental Health Service since 1986. Dr. Kern also held an appointment as associate professor at the Brown University Medical School in Providence, although his salary came exclusively from the hospital. In 1994, a worker at the Microfibres plant (of Pawtucket, Rhode Island) developed shortness of breath and a cough; he was ultimately referred to Kern for diagnosis. The manufacturing process used at all Microfibres plants (there was also a plant in Kingston, Ontario) involves gluing finely cut nylon filament (known as flock) onto cotton and polyester fabric. The process produces a velour-type fabric that is used as a seat covering for furniture and automobiles.

After signing a confidentiality agreement, Kern visited the plant but could find no cause of the patient's condition. A year later, when a second worker was referred to him with similar symptoms, Kern contacted the company suggesting that something at the plant may explain the illness of the two workers. He recommended that the company hire him through Memorial Hospital to evaluate whether there were health hazards in the plant; he also recommended that the company contact the National Institute of Occupational Safety and Health (NIOSH). Subsequently, Microfibres reported the incident to NIOSH and, in an oral agreement, hired Kern to investigate the possible causes of the mysterious lung illness.[36] Eventually, it became known as "flock worker's lung." Other cases were in fact reported in the early 1990s at the Kingston, Ontario, plants.

While the medical investigation was still ongoing in October 1996, Kern prepared an abstract on the illnesses he observed for submission to a 1997 meeting of the American Thoracic Society.[37] His purpose in writing the abstract was twofold: to warn other occupational physicians about the disease and to learn from their own experiences. Microfibres opposed his publication of the abstract on the basis that his conclusions were premature and that he had signed a confidentiality agreement when he visited the plant in 1994.[38] Kern did not believe the agreement he signed covered the reporting of an occupational disease that he discovered at the plant, but Microfibres did consider the agreement binding to the work that he did, which was paid for by the company. Kern was willing to compromise by rewriting his abstract so as not to mention the company or its locations. Microfibres rejected the compromise, but Kern decided to publish the abstract nevertheless. At Brown University, administrators had differing views about whether Kern had a legal right to publish the abstract.[39] A university associate dean urged Kern to cancel the presentation of his findings in the Microfibres investigation, and Kern demurred.

One week after he presented his results at the meeting of the American Thoracic Society in May 1997, Kern received a letter from Memorial Hospital and Brown University (the two appointments were linked) informing him that his five-year employment contract would not be renewed after its expiration in 1999. The hospital claimed that his dismissal was not in retaliation for his submitting the abstract.

In most media accounts, the Kern case was cast as a conflict between the responsibility of a clinical researcher who is hired by a company to evaluate potential occupational hazards (i.e., to inform) and a company that has a right to protect its proprietary interest. Why was the contract of the clinical investigator terminated so soon after he released his findings? One reason given was that the hospital was fearful of a lawsuit. Another reason was suggested by Kern himself. He claimed that the family that owns Microfibres contributed substantially to the hospital and that members of that family serve as members of the hospital's corporation.[40] Kern was quoted in the *Boston Globe*: "In the end, when the smoke clears, the lesson is that scientific findings can be suppressed if there are financial interests, and there will be little response from the university, whether or not there are public health implications."[41]

Like the Kern case, the next case involves a physician, a university, and a hospital under conditions where corporate leverage played a role in the dismissal (albeit temporarily) of a clinical investigator from the directorship of a medical program.

Dr. Nancy Olivieri is a hematologist who specializes in the treatment of hereditary blood diseases—especially thalassemia, where the body does not produce normal blood cells. She had appointments at the University of Toronto and one of its associated teaching hospitals, the Hospital for Sick Children (HSC). Olivieri's thalassemia patients required routine blood transfusions to replace abnormal blood cells. The transfusions themselves eventually introduce risk, which arises from the deposition of excess iron in the body's major organs. Too much iron can be fatal, so it therefore has to be removed. Olivieri became interested in studying an experimental iron-chelation drug called deferiprone, which looked promising at the time for reducing the buildup of iron.

In 1993, Olivieri entered into a collaboration with the Canadian drug manufacturer Apotex to conduct a randomized trial. The purpose of the trial was to compare its drug with the standardized treatment for the disease, a drug called deferoxamine. The contract Olivieri signed contained a confidentiality clause giving the sponsor of the research the sole right to communicate the trial data for one year after the termination of the trial. This provision of the contract was in full accordance with the university's policy on external research contracts. Meanwhile, the university and Apotex were engaged in discussions about a substantial donation for a new biomedical research center to be located on the campus. A nonbinding agreement was reached in 1998 under which Apotex would donate $12.7 million to finance a research center at the University of Toronto. A potential gift of this magnitude would make any university exquisitely attentive to the interests of the donor. In fact, the donor made the gift contingent on the university's lobbying the Canadian government to delay regulations that were unfavorable to the generic drug industry.[42] The initial results from the new chelation drug in reducing or maintaining liver iron levels in thalassemia patients were encouraging. Questions were raised, however, about the drug's efficacy. To answer those questions Olivieri began a second trial, but this time she did not sign a confidentiality agreement with Apotex.[43]

In 1996, when Olivieri was engaged in the second trial, she observed that the drug began losing its efficacy, and in some cases, it exposed her patients to additional risk. Apotex, which disputed her interpretation of the risks, threatened legal action against Olivieri for violating the conflict-of-interest agreement of the first trial if she released information about the second trial to her colleagues or her patients. The company also blamed Olivieri for protocol violations and terminated the trial that she was supervising.

The initial contract signed by Olivieri and approved by the University of Toronto and its affiliated Hospital for Sick Children was explicit about the company's right to suppress information during the trial and for one year after its termination. The hospital's research ethics board (REB) did not require any safety clauses in the protocol that would protect the interests of trial participants in the event that there were severe adverse side effects. Typically, such clauses provide an override to the one-year gag order on communication to conform to the physician's duty to inform. The chair of the research ethics board agreed that Olivieri should inform participants about the risks of the drug through consent forms, a recommendation at odds with the company's policy. It was after Olivieri submitted a revised consent form to the REB in May 1996 that the company terminated the clinical trials.

Despite the warnings Olivieri received from Apotex of legal actions that the company might take against her should she communicate risks to patients or anyone else without prior permission of the company, the physician accepted her "duty to inform" as preemptive of contractual obligations. She published two abstracts based on the data she collected from the discontinued trials, presenting those results at a 1997 conference.

After publishing the abstracts, Olivieri was removed from the directorship of HSC's hemoglobinopathy program. Subsequent to Olivieri's removal as director, the HSC and the university assumed responsibility for her legal defense against Apotex, affirmed her right to academic freedom, and acknowledged false allegations made against her. In 1998, the university and Apotex agreed to suspend discussions about its multimillion-dollar gift until the dispute involving Olivieri was resolved. A year later, the company withdrew from its 1998 agreement. In October 2001, a committee of three professors ruled that Nancy Olivieri behaved ethically and professionally.[44] She was subsequently reappointed to her position. Other independent reviews of the case by the Canadian Association of University Teachers and the College of Physicians and Surgeons also vindicated Olivieri of any wrongdoing.

This case has brought attention to the problem of university conflicts of interest, which complicate efforts by faculty to exercise moral leadership and which create institutional sources of bias. In March 2001, the University of Toronto and its affiliated teaching hospitals adopted a new policy that would not permit contracts with clauses that prevented clinical investigators from disclosing risks to patients. According to a recent study, however, restrictive clauses of this nature are still accepted by many institutions in the United States and Canada.[45]

UNIVERSITY CONFLICT OF INTEREST: A SEARCH FOR STANDARDS

The rash of media attention on corporate–university ties has had a curious effect. On one hand, it has raised the awareness of administrators that they are facing a public relations issue. However, as long as the relationships do not spawn new regulations or create losses of government funding, administrators have little reason to set higher ethical standards for faculty. On the other hand, it has hardened the position of some universities who decry any standards that restrict private funding. As a matter of record, easing the ethical guidelines has improved the flow of private sector money to the university. I can remember the words of the former president of Tufts University, Jean Mayer, who took a principled stand against accepting weapons research; however, he deftly compromised his position for the missile-defense shield ("Star Wars") program. He used to say jokingly, "The only thing wrong with tainted money is there t'aint enough of it." Since the norms of research are evolving, many universities do not want to be left behind while others are reaping the rewards for lowering the conflict-of-interest standards.

Since 1988, after its infamous Tseng case (see chapter 2), Harvard Medical School set relatively high standards for its academic faculty. Medical faculty members cannot conduct research for a company in which they own more than $20,000 in stock or receive more than $10,000 in consulting fees or royalties, a policy that covers more than 6,000 full-time faculty. Also, faculty are not permitted to spend more than 20 percent of their time (e.g., one day a week) working outside the university, a norm on the books but rarely enforced in many institutions. When publishing articles or findings, faculty must disclose any financial interests they have in the companies that support the research or that have intellectual property rights to it. In 2000, Harvard contemplated reducing the restrictions putatively to make the university more competitive. Their medical researchers, a Harvard dean argued, were willing to leave for other schools where they could improve their personal incomes by consulting and through lucrative equity arrangements. In a report entitled *Guidelines for Research Projects Undertaken in Cooperation with Industry,* Harvard's Standing Committee on Research Policy proposed a new set of rules that permitted greater flexibility in cooperative agreements and that increased not only the amount of stock faculty were permitted to hold but also the consulting money they could earn from companies that sponsor their research.

In contrast to Harvard's conservative position on conflicts of interest, Stanford University School of Medicine does not set fixed limits on stock ownership or royalties. Faculty members who own more than $100,000 worth of stock or 0.5 percent of a company must notify the university, which then decides on a case-by-case basis whether any restrictions should be applied. At MIT, the operative rule has more to do with whether a faculty member's equity in a company is large enough to influence the company's stock rather than the dollar value of the equity.

In May 2000, Harvard reconsidered its course and decided it would not pursue the proposal for easing its conflict-of-interest guidelines. This decision came in the wake of a widely publicized case at the University of Pennsylvania involving the death of a teenager who was enrolled in a clinical trial of a gene therapy experiment. During an investigation, it was revealed that the principal investigator of that trial had commercial ties to a private company, ties that were not disclosed to the subject.

What is the current state of university conflict-of-interest guidelines? What types of incentives are there for universities to develop meaningful and reliable guidelines? A national survey taken of 127 medical schools and 170 other research institutions that receive over $5 million in total grants annually from the National Institutes of Health (NIH) and the National Science Foundation (NSF) provides some useful information about the status and diversity of conflict-of-interest policies within academia.[46] With an 85 percent response rate, the researchers had a survey sample of 250 medical schools and other research institutions. Fourteen of the respondents reported that they had no policies on conflicts of interest (COI). Of those that did have policies, 92 percent had become effective after June 28, 1994, which marked the date the first federal draft guidelines on COI were released. Federal mandates on COI guidelines are the most important incentive for academic research institutions to adopt guidelines and implement management procedures.

The study found significant variations in university guidelines, but they all had one common theme: "that the management of conflicts and the penalties for nondisclosure were totally discretionary."[47] Only one institution reported that it had a mandatory strategy for managing the initial disclosure of conflicts of interest. A General Accounting Office (GAO) report released in November 2001 confirmed what many observers already suspected—namely, that there are significant deficiencies in the way that conflicts of interests are handled at universities.[48] The GAO did an in-depth study of five major research universities, among the twenty institutions that receive the largest NIH funding for biomedical research. This represents about 1 percent of all the institutions with NIH funds. The most damning of the GAO results was that universities allow investigators to determine whether they are in compliance with the institution's conflict-of-interest policies and that monitoring of faculty activities is practically nonexistent. Managing conflicts of interest appears to be a euphemism at most institutions making a superficial attempt to avoid any embarrassment from the entrepreneurial practices of faculty.

Faculty conflicts of interests are not the only challenge facing universities. Academic institutions and nonprofit research centers are increasingly becoming equity partners in for-profit ventures, and they also often share intellectual property rights over patents from faculty discoveries. In other cases, high-level university administrators—including presidents, provosts, and deans—serve on the board of directors of companies that have a relationship with the university. When such relationships place the university in compromising roles or when they give the appearance that the university has a financial interest in the outcome of certain research programs, this scenario is usually termed an "institutional conflict of interest." Some argue that institutional conflicts of interest are a greater source of concern than those associated with individuals. Resnik and Shamoo note: "Since institutional COIs can affect the conduct of dozens or even thousands of people inside and outside the institution, they have a potentially greater impact than individual COIs."[49] Ironically, with all the discussion in the media about individual conflicts of interest, there has been little attention given to the institutional conflicts.

When the Department of Health and Human Services (DHHS) was preparing its conflict-of-interest regulations for universities and its grantees in 1994, it considered the issue of institutional conflicts. The agency collected commentaries on its proposed regulations and noted that there was near unanimity that its newly fashioned regulations should not address institutional conflicts of interest. The DHHS agreed with that recommendation and excluded from its 1995 final regulations any discussion of institutional conflicts of interest, which it said would be considered in a separate process.[50] To date, there has not been a separate process. Under its regulations, universities are required to manage the conflicts of interest of its faculty. But who will manage the institution's conflicting financial interests? The dilemma of managing institutional COIs was highlighted by Resnik and Shamoo:

> Universities have no one who has credible, moral authority to oversee or manage their relationships with private corporations. To whom would a university disclose its COIs? Who would decide whether these COIs can be managed and monitored or should be prohibited or avoided? While formal governing bodies, such [as] boards of trustees or regents, have the legal authority to oversee these relationships, they themselves may be in the middle of the very conflicts they would be asked to manage.[51]

A case in point is that of Boston University, which maintained a close connection with its spin-off company Seragen—a pharmaceutical start-up. Blumenthal reported that in 1994 "the university itself, individual members of its board of trustees, the president of the university, and members of the its faculty own substantial equity in this company, which is also funding research at the university."[52] It doesn't get more conflictual than that.

Suppose a medical school has equity in a for-profit company that performs clinical trials for drug companies. Should that relationship be taken into consideration when a research center at the university applies for a federal grant to study a drug that is part of the clinical trial? When are the university's financial interests in direct conflict with the research that gets performed within its facilities? Can researchers on the university payroll engage in disinterested science when the outcome may have substantial impacts on the institution's endowment? Should researchers ever be placed in that situation?

When universities and their faculty become partners in commercializing research, conflicts are inescapable. One study of university entrepreneurship asked: "Who is the fiduciary when universities convert a professor's discovery into equity ownership in a company, and that company is subsequently sued for patent infringement?"[53] The authors site a case where a faculty member filed a lawsuit against his university for licensing his discovery at discount rates.

Academic research milieu is generally acknowledged to be permanently and unabashedly linked to the private sector. No responsible voices have called for an end to corporate sponsorship of research. Most responses to managing conflicts of interest are focused on writing responsible contracts, setting restrictions on faculty with equity interests in a company, and ensuring that investigators are fully in control of the research regime, from framing the research design to publishing the results.

But are there some corporate sponsors that should be denied access to the university, even if their motives appear philanthropic? This question was one that administrators at the University of Nottingham confronted in December 2000. The university was offered a $5.3 million gift from British American Tobacco for the establishment of Britain's first "International Center for Corporate Social Responsibility."[54] At the time of its gift, the tobacco company was enmeshed in litigation on the health effects of cigarettes. The irony of accepting funding from a tobacco conglomerate for a center on social responsibility was not missed by members of the university community. A report of a World Health Organization panel, based on confidential documents obtained during litigation, showed that tobacco companies set up front scientific organizations and funded advocacy science to dispute responsible studies linking tobacco to cancer.[55] Could the university's association with such companies compromise or blemish its integrity? Would such funding establish a patronage relationship between big tobacco and the University of Nottingham that could spill over to tobacco research and to the institution's ethical behavior? The university defended its decision to keep the tobacco funds, arguing that it was legal and that the institute it creates would be fully independent and under strict university ethics guidelines. Furthermore, university administrators argued, the money the center spends would benefit society.

Within a year of the award, one funding source withdrew over $2 million in research grants supporting work at the University of Nottingham and transferred it to another university. The editor of the prestigious *British Medical Journal* resigned his adjunct professorship at the university in response to the tobacco company gift.

Most of the incentive structure for universities weighs in favor of turning a blind eye toward faculty and institutional conflicts of interest. Unless all parties agree to some form of reasonable accountability, transparency, and sanctions, the use of guidelines to turn the tide away from institutional complicity and neglect of the problem is unlikely to succeed. Universities, once considered a neutral party in conflicts among stakeholder groups, now join the ranks of those groups. Writing in the *Chronicle of Higher Education*, Raymond Orbach, chancellor and professor of physics at the University of California,

Riverside, argued that universities can serve as "honest brokers" between business and the public sector in public policy controversies, in areas such as environmental protection. What Orbach failed to see is that the "honest broker" model of universities cannot work in the current climate. He wrote, "Finding a way to avoid even the appearance of a conflict is one of the principal challenges in establishing the university as an honest broker."[56] Of note is the fact that one of the sister University-of-California schools (University of California, Irvine) was caught up in a scandal involving the close relationships its cancer researchers had with business.[57]

Few, if any, universities can claim such neutrality. The field of biotechnology has notably been replete with such conflicts. A 1992 editorial in *Nature* stated: "Ever since academic biologists took their first tentative steps toward industry in the late 1979s and early 1980s, when companies such as Hoechst, DuPont, and Monsanto were contributing millions of dollars to university collaborations, the research community has worried about its academic soul."[58] And what would the soul of academia be without the pure virtue of the pursuit of knowledge and the protection of that pursuit from commodification and distortion by the marketplace? The next chapter looks at how the concept of "intellectual property" evolved from its origins in technique to its current incarnation as scientific knowledge. These changes have turned universities into knowledge brokers in a new marketplace for scientific discovery. The rapid commercialization taking hold at universities was significantly facilitated by new rules on intellectual property.

NOTES

1. Roger L. Geiger, *Research and Relevant Knowledge* (Oxford: Oxford University Press, 1993).

2. Francis X. Sutton, "The Distinction and Durability of American Research Universities," in *The Research Universities in a Time of Discontent*, ed. J. R. Cole, E. G. Barber, and S. R. Graubard (Baltimore: The Johns Hopkins University Press, 1994).

3. Geiger, *Research and Relevant Knowledge*, 299–300.

4. James Ridgeway, *The Closed Corporation* (New York: Random House, 1968), 84.

5. Robert M. Rosenzweig, *The Research Universities and Their Patrons* (Berkeley: University of California Press, 1992), 42.

6. Rosenzweig, *The Research Universities,* 59.

7. Rosenzweig, *The Research Universities,* 43.

8. G. A. Keyworth II, "Federal R&D and Industrial Policy," *Science* 210 (June 10, 1993): 1122–1125.

9. Wesley Cohen, Richard Florida, and W. Richard Goe, *University–Industry Research Centers.* A report of the Center for Economic Development, H. John Heinz III School of Public Policy and Management, Carnegie Mellon University, July 1994.

10. U.S. Congress, Office of Technology Assessment (OTA), *New Developments in Biotechnology: 4. U.S. Investment in Biotechnology* (Washington, D.C.: USGPO, July 1988), 115.

11. OTA, *New Developments,* 114.

12. OTA, *New Developments,* 113.

13. National Research Council, introduction to *Intellectual Property Rights and Research Tools in Molecular Biology,* a summary of a workshop held at the National Academy of Sciences, February 15–16, 1996 (Washington, D.C.: National Academy Press, 1997), 1.

14. Sheila Slaughter and Larry L. Leslie, *Academic Capitalism* (Baltimore: The Johns Hopkins University Press, 1997), 223.

15. U.S. Congress, House Subcommittee on Investigations and Oversight; House Subcommittee on Science, Research and Technology; Committee on Science and Technology, *University/Industry Cooperation in Biotechnology,* 97th Cong., June 16–17, 1982 (Washington D.C.: USGPO, 1982), 2.

16. U.S. Congress, *University/Industry Cooperation,* 6.

17. U.S. Congress, House Committee on Government Operations, Subcommittee on Human Resoruces and Intergovernmental Relations, *Are Scientific Misconduct and Conflicts of Interest Hazardous to Our Health?* (Washington, D.C.: USGPO, September 10, 1990), 68.

18. Maura Lerner and Joe Rigert, "Audits say 'U' Knew of ALG Problems," *Star Tribune,* August 23, 1992, 1B, 3B.

19. Maura Lcrncr and Joc Rigert, ""U' is Forced to Halt Sales of Drug ALG," *Star Tribune,* August 23, 1992, A1, 6–7.

20. Joe Rigert and Maura Lerner, "Najarian Admits Mistakes were Made in ALG Drug Program," *Star Tribune,* August 26, 1992, B1, B2.

21. Goldie Blumenstyk, "Berkeley Pact with a Swiss Company Takes Technology Transfer to a New Level," *Chronicle of Higher Education*, December 11, 1998, A56.

22. Will Evans, "UC-Berkeley Alliance with Novartis Blasted by State Senators," *Daily California,* May 16, 2000.

23. J. Walsh, "Universities: Industry Links Raise Conflict of Interest Issue," *Science* 164 (April 25, 1969): 412.

24. Miguel Altieri (talk to the Environmental Grantmakers Association, New Palz, N.Y., September 13, 2000).

25. Ignacio Chapela and David Quist, "Transgenic DNA Introgressed into Traditional Maize Landraces in Oaxaca, Mexico," *Nature* 414 (November 29, 2001): 541–543.

26. Nick Kaplinsky, David Braun, Damon Lisch et al., "Maize Transgene Results in Mexico Are Artifacts," *Nature* 415 (April 4, 2002): 2.

27. Paul Elias, "Corn Study Spurs Debate on Links of Firms, Colleges," *Boston Globe,* April 22, 2002, D6.

28. Kelsey Demmon and Amanda Paul, "Mexican Investigation Validates Corn Study," *The Daily Californian,* August 20, 2002.

29. Public Citizen, *Safeguards at Risk: John Graham and Corporate America's Back Door to the Bush White House* (Washington, D.C.: Public Citizen, March 2001).

30. Sheldon Krimsky, *Hormonal Chaos: The Scientific and Social Origins of the Environmental Endocrine Hypothesis* (Baltimore: The Johns Hopkins University Press, 2000).

31. Douglas Jehl, "Regulations Czar Prefers New Path," *New York Times*, March 25, 2001, A1, A22.

32. Dick Durbin, "Graham Flunks the Cost-Benefit Test," *Washington Post,* July 16, 2001, A15.

33. Durbin, "Graham Flunks," A15.

34. Julie Smyth, "Psychiatrist Denied Job Sues U of T: Linked Prozac to Suicide," *National Post,* September 25, 2001.

35. Nicholas Keung, "MD Settles Lawsuit with U of T over Job," *Toronto Star,* May 1, 2002, A23.

36. Donald J. Marsh, "A Letter from Dean Marsh Regarding Issue of Academic Freedom," *George Street Journal* [Brown University publication] 21, no. 30 (May 23–29, 1997), B13.

37. Miriam Schuchman, "Secrecy in Science: The Flock Worker's Lung Investigation," *Annals of Internal Medicine* 129 (August 15, 1998): 341–344.

38. Wade Roush, "Secrecy Dispute Pits Brown Researcher against Company," *Science* 276 (April 25, 1977): 523–524.

39. Marsh, "A Letter from Dean Marsh," p. 4.

40. Schuchman, "Secrecy in Science."

41. Richard A. Knox, "Brown Public Health Researcher Fired," *Boston Sunday Globe,* July 6, 1997, B1, B5.

42. Bernard Simon, "Private Sector: The Good, the Bad and the Generic," *New York Times,* October 28, 2001, sec. 3, p. 2.

43. David G. Nathan and David J. Weatherall, "Academic Freedom in Clinical Research. Sounding Board," *New England Journal of Medicine* 347 (October 24, 2002): 1368–1369.

44. Editorial, "Keeping Research Pure," *Toronto Star,* October 29, 2001, A20.

45. Patricia Baird, Jocelyn Downie, and Jon Thompson, "Clinical Trials and Industry," *Science* 297 (September 27, 2002): 2211.

46. S. van McCrary, Cheryl B. Anderson, Jolen Khan Jakovljevic et al., "A National Survey of Policies on Disclosure of Conflicts of Interest in Biomedcal Research," *New England Journal of Medicine* 343 (November 30, 2000): 1621–1626.

47. S. van McCrary, "A National Survey of Policies," p. 1622.

48. U.S. General Accounting Office, *Biomedical Research: HHS Direction Needed to Address Financial Conflicts of Interest.* Report to the ranking minority member, Sub-

committee on Public Health, Committee on Health, Education, Labor, and Pensions (Washington, D.C.: GAO, November 2001) GAO-02-89.

49. David B. Resnik and Adil E. Shamoo, "Conflict of Interest and the University," *Accountability in Research* 9 (January–March 2002): 45–64.

50. Department of Health and Human Services, Public Health Service, "Objectivity in Research," *Federal Register* 60 (July 11, 1995): 35810–35819.

51. Resnik and Shamoo, "Conflict of Interest and the University," 56.

52. David Blumenthal, "Growing Pains for New Academic/Industry Relationships," *Health Affairs* 13 (Summer 1994): 176–193.

53. Walter W. Powell and Jason Owen-Smith, "Universities as Creators and Retailers of Intellectual Property: Life-Sciences Research and Commercial Development," in *To Profit or Not to Profit,* ed. Burton A. Weisbrod (Cambridge, U.K.: Cambridge University Press, 1998), 190.

54. Dan Ferber, "Is Corporate Funding Steering Research Institutions Off Track?" *The Scientist,* February 5, 2002.

55. Committee of Experts on Tobacco Industry Documents, World Health Organization, *Tobacco Company Strategies to Undermine Tobacco Control Activities at the World Health Organization,* July 2000, at filestore.who.int/~who/home/tobacco/tobacco.pdf (accessed March 1, 2002).

56. Raymond L. Orbach, "Universities Should Be 'Honest Brokers' between Business and the Public Sector," *Chronicle of Higher Education,* April 6, 2001, 13.

57. Jeff Gottlieb, "UCI Case Raises Issue of Schools' Ties to Business," *Los Angeles Times,* December 27, 1998, A28.

58. Editorial, "Conflict of Interest Revisited," *Nature* 355 (February 27, 1992): 751.

4

KNOWLEDGE AS PROPERTY

In February 2000, the U.S. Patent and Trademark Office issued Human Genome Sciences, a Maryland biotechnology company, a patent on a gene that was touted as possibly providing the clue to how the AIDS virus infects human cells. Members of the company discovered and sequenced the gene that codes for a receptor molecule (called CCR5) that allows the HIV virus to gain entry into the cells. Individuals who have a mutation of the gene and therefore do not produce a clean copy of the protein appear not to contract AIDS when exposed to the virus.[1] The discovery of the gene may give scientists a clue to disabling the receptor that allows HIV to infect human cells. When Human Genome Sciences applied for the patent on the gene around June 1995, it did not say anything in the patent application about the protein's role as a pathway for HIV virus to enter cells. It did discuss the possibility that the gene or the protein it synthesizes could have an impact on a variety of diseases and that it could function as a viral receptor. By July 2000, Human Genome Sciences had been awarded more than one hundred human gene patents with 7,500 patent applications pending. The patent on CCRS gives the company a twenty-year monopoly over the use of the AIDS receptor gene and its protein for developing any assays or therapeutic drugs for diseases including drugs to counteract AIDS.[2]

Many people have been perplexed by the practice of awarding patents for the genes of living things. Aren't those natural substances? Isn't the patent system designed to provide incentives for inventions? The company did not invent the gene; rather, the gene is part of the natural evolutionary heritage of Homo sapiens. What is really curious is that the patent was awarded even before the company knew the gene was suspected of having anything to do with AIDS and certainly before any therapeutic use was developed. How can patents be given for "natural substances" like genes? Once the gene is patented, the patent holder can restrict the use of the gene sequence to researchers or companies who pay a licensing fee to use it. The patent holder could offer restrictive licenses to a preferred client—thus keeping the number of competitors for finding a cure for AIDS to a limited group—or the company may offer nonrestrictive licenses, which would provide for greater competition.

HISTORICAL ROOTS OF INTELLECTUAL PROPERTY

The idea behind protecting intellectual property, which is the economic value that can be derived from one's creative writing or invention, is neither uniquely American nor capitalist in origin. Historians have records of protected inventions in feudal societies dating back to the Middle Ages. The city of Venice granted ten-year monopolies to inventors of silk-making instruments in the 1200s. One of its most notable patent recipients was the physicist Galileo, who was granted his patent by the Venetian Senate in 1594 for his invention of a horse-driven water pump. In 1624, England passed the Statute of Monopolies, which gave inventors the right to obtain patents on their inventions.

The United States has extended the concept of intellectual property and patents to entities that our forefathers could not have even imagined and that many contemporaries find perplexing as well. The U.S. Patent and Trademark Office (USPTO) has awarded patents for animals, DNA, genes, microorganisms, plants, and chemicals found in nature. How did we get from patenting the cotton gin and Pasteur's microbe for making wine to patenting a gene for AIDS? When is knowledge patentable? How has the "gold rush" for DNA prospecting of the lucrative gene sequences in the human genome affected the culture and practice of biomedical science?

Among the Founding Fathers of the United States, Benjamin Franklin and Thomas Jefferson were distinguished inventors. Jefferson, in particular, had given serious thought to how the new Constitution should promote science and invention. Prior to 1790, the American colonies granted patents by a special act of the colonial legislatures. The governing body of each colony protected the rights of inventors, yet Jefferson still had some initial reservations about including a patent provision in the Constitution. Jefferson wanted to have a bill of rights included in the Constitution to guarantee free speech, freedom of religion, and protection against monopolies—but a patent was a kind of monopoly, at least for a period of years.

James Madison, on the other hand, believed that patents were a just sacrifice for their benefit in stimulating the practical arts. He wrote to Jefferson in 1788:

"With regard to monopolies they are justly classified among the great nuisances in government. But it is clear that as encouragements to literary works and ingenious discoveries, they are not too valuable to be wholly renounced."[3] In the *Federalist Paper* #43, Madison wrote: "The copyright of authors has been solemnly adjudged in Great Britain to be a right of common law. The right to useful inventions seems with equal reason to belong to the inventors."[4] Jefferson eventually agreed to permit monopolies for copyrights and inventions for a fixed number of years; interestingly, he never took out patents on his own inventions. He believed that human ingenuity, expressed through invention, should not be privately owned. He also believed that inventors were given entitlement through a social contract with society to the profits arising from the invention, with the proviso that they disclose the secret behind their discovery. Ultimately, however, he came around to accepting Madison's argument that patents could encourage human ingenuity.

The rest is history. The right to a copyright or a patent was the only fundamental right included in the body of the Constitution, and the Bill of Rights was adopted as a separate document. Article 1, section 8 of the Constitution states: "The Congress shall have the power to promote the progress of science and useful arts by securing for a limited time to authors and inventors the exclusive right to their respective writings and discoveries." The Patent Act of 1790 established the U.S. Patent Office. Patent applications were reviewed by a three-person board, consisting of the secretary of state (Jefferson), the secretary of war, and the attorney general. During a three-year period, the patent board reviewed 114 applications and granted forty-nine patents. In 1793, Congress passed another patent act, which created a special patent office, relieving the cabinet officers of the task of patent review.

AMERICAN PATENT POLICY

For over two hundred years, the U.S. Patent and Trademark Office has been greasing the wheels of progress with patents that have taken us through the industrial revolution, the chemical age, and now the revolutions in computer science and biotechnology. According to the U.S. Patent Act (35 U.S.C. 101), a patent may be awarded to "whoever invents or discovers any new and useful process, machine, manufacture, or composition of matter, or any new and useful improvement thereof." Throughout this period, the patent laws and their interpretation have given rise to patents on living organisms, natural substances, and the fundamental unit of heritability—the gene. How do we reconcile the idea that patents are awards for invention and that they are given for objects of nature? The key to understanding this idea is that to be considered patentable matter, some human intervention must have transformed the natural object from one state to another to give the object its social utility. To receive a patent, the applicant must demonstrate that the "invention" has utility, novelty, and nonobviousness. In 1912, the U.S. Supreme Court upheld a patent on an isolated and purified form of adrenaline. It was considered novel over its natural form. Thus, you can patent a natural substance if it requires some ingenuity and inventiveness to purify it or deliver it from its obscured state. We would not ordinarily consider a natural object that has been isolated from nature a "product of manufacture" or a "new composition of matter," but the courts have given the USPTO the mandate to patent such objects. Patents have been awarded for microorganisms that have been isolated from soil. We can therefore state one of the central principles in patent law and regulation as follows: Chemical and biological compositions of matter are patentable if through human ingenuity they are put into a form in which they do not exist in nature. For example, in his study of life patents, Louis Guenin noted: "The U.S. Patent and Trademark Office . . . issued a patent to Glenn Seaborg on curium and isotopes of americium, transuranic elements believed to exist on earth only in a cyclotron or reactor as a result of human efforts."[5] Likewise, the first patent for isolating nucleic acid was awarded in 1945, and twenty-one years later a patent was given for preparing ribonucleic acid by a fermentation process.[6]

So how does that principle apply to genes? Gene patenting is treated like patenting new chemical compounds. Two grounds are used to justify gene patents. The first is the "needle in the haystack" principle. In other words, some argue that it takes ingenuity to find and isolate the genes within the genome. The pure gene, removed from its chromosome, does not occur in nature—at least, not in any obvious way. Once isolated and purified, the gene can be cloned (reproduced) in another organism and used to produce an assay that can detect the gene or some mutant form of it. The second justification for gene patents falls under the idea of "new compositions of matter." For this rationale, consider the argument that appeared in a letter to the journal *Nature,* where a commentator describes the legal foundations for patenting genes:

> Antibiotics have been patented for years with the challenges that have been applied to DNA sequences, but they are also natural molecules produced by living organisms and are, in a very real sense, discovered. Why, then are they classified as inventions and therefore patentable? The fact is that the antibiotics produced by living organisms in their wild state are not patentable. However, if an individual or a company searches for and finds the right organism, identifies a useful property of one of its metabolites, isolates, purifies and characterizes the product

> and devises a method of making and using it, then the outcome is not an antibiotic in its natural state, but the genuine product of human ingenuity.[7]

The author draws an analogy between naturally extracted and purified antibiotics and DNA: "In their natural state, within the body of an organism, directing the production of enzymes, hormones or whatever, they are not patentable. But a DNA sequence that can specify erythropoietin, for example, when identified, characterized and transferred to a suitable production . . . is justifiably patentable."[8]

Scientists are not seeking patent applications for excised DNA from a human chromosome in its natural state. With the mechanized processes that exist for gene sequencing, which have become part of routine practice in genomic work, construction of DNA sequences may not by itself meet the novelty criteria. However, scientists transform the DNA sequence into a version called "complimentary DNA" (or cDNA). A gene that codes a protein typically has many redundant or irrelevant nucleotides, which are not essential for the protein synthesis. When the extraneous sequences (introns) are removed, the version that is left is cDNA. Since cDNA is not found in nature, it has been argued that the patented entity is therefore not a product of nature. This point was explained by Kevles and Hood in their book *The Code of Codes:*

> Genes are products of nature, at least as they occur with their introns and exons in the cellular chromosomes. However, the so-called copy DNA (cDNA) version of the gene, with the introns edited out, does not occur naturally. It is coded into messenger RNA by the process that reads the raw cellular DNA, but it is not physically realized in the cell. Since it can be physically realized by a devising of human beings, using the enzyme reverse transcriptase, it is patentable.[9]

LIFE PATENTS THEN AND NOW

We are still talking about patenting entities, not physical laws or abstract ideas; thus, we cannot patent the fundamental principles of genetics. But is there a clear demarcation between patenting entities and patenting knowledge? And what about the distinction between patenting inert entities and patenting living things?

The issue of whether the U.S. patenting law reveals a congressional intent to award patents for living things in and of themselves was the subject of the 1980 Supreme Court decision in *Diamond v. Chakrabarty.* Prior to that decision, plenty of patents were issued for microorganisms. In 1873, Louis Pasteur was awarded U.S. patent number 141,072 for a strain of yeast, the first of several patents for life forms; however, the patent claim involved the organism within a process, not just the organism itself. No one could claim monopoly control over the organism independently of how it was used in an invention—that is, until the 1980 Supreme Court decision.

In *Diamond v. Chakrabarty*,[10] a researcher (Dr. Ananda Chakrabarty) working for General Electric selected a soil microorganism called *Pseudomonas*, which possessed certain properties for degrading hydrocarbons. He then mixed together plasmids (circular pieces of DNA, each with specific hydrocarbon metabolizing capacities) from several *Pseudomonas* varieties into his choice strain. The patent claim, filed June 7, 1972, was thus twofold: one, for a process of degrading oil from oil spills using the "novel" *Pseudomonas* strain; and two, for the strain itself. The abstract of the patent application stated: "Unique microorganisms have been developed by the application of genetic engineering techniques. These microorganisms contain at least two stable (compatible) energy-generating plasmids, these plasmids specifying separate degradative pathways."[11] According to the claim, each of the plasmids provide a separate degradative pathway for hydrocarbons, giving the organism its special role in breaking down oil from oil spills. The patent application contained thirty-six claims that fall into three categories: first, the methods of producing the bacteria (the process); second, the inoculum, which comprised materials like straw that would carry the bacteria by floating on water; third, the bacterium itself. The patent examiner allowed the claims for the process and the inoculum but rejected the claim for the bacterium as a patentable entity. The patent office rejected the third claim, notwithstanding the inventor's modifications, because it held that microorganisms are both products of nature and living things, which preclude them from being patented.

Of course, patents have been awarded for natural products if they were appropriately modified or selected from nature. In this case, the oil-digesting *Pseudomonas* bacterium was manipulated to receive DNA from other sources and thus not found in nature. The USPTO Board of Appeals overturned the patent office's decision that the bacterium was a product of nature but upheld the patent rejection on the grounds that living things were not, in themselves, patentable subject matter. As long as they were part of a process, they could be patented. The Court of Customs and Patent Appeals reversed the Board of Appeals' decision and said that it was not of legal significance for patent law, whether the entity was living or nonliving matter. Finally, in a five-four ruling, the U.S. Supreme Court affirmed the decision of the highest appeals court that a living, human-modified microorganism is patentable subject matter under section 101 of the patent law as a "manufacture" or "composition of matter."

On what basis did the majority of five Supreme Court justices render its decision? What was the minority argument?

The judges searched the patent legislation to locate any explicit language that could provide a clue to congressional intent with regard to the patenting of living things. During the 1920s, plant breeders lobbied Congress to gain the benefits of patenting that had been granted to innovators in the mechanical arts. In response, Congress passed the Plant Patent Act of 1930, which extended the definition of patentable material to certain varieties of asexually produced plants. These plants were propagated by cuttings, grafting, and budding—but not by seeds. There remained no patent rights over seeds, even if they were altered by human artifact such as radiation. Supporting documentation from the House and Senate revealed congressional intent that, from a patenting standpoint, there was no distinction between plant grafters and chemists who create new compositions of matter.

So Congress did not have a principled objection to the patenting of life forms—and neither did the public for that matter. A mere three months passed between the date the act was introduced to the date it was passed. Nevertheless, it was deemed a congressional decision to extend the laws of intellectual property to certain plant varieties. Forty years later, Congress revisited plant patenting, but this time for sexually reproduced plants.

The 1960s had seen a revolution in hybrid seed development. Hybridization techniques involve an arduous and time-consuming process of crossing, screening, and selecting new strains of seeds with desirable traits, such as wind and salt tolerance. The development and deployment of hybrid strains of corn and rice became known as the "Green Revolution." Seed manufacturers sought patent protection for their varieties. Congress again responded, this time by passing the Plant Variety Protection Act of 1970, which brought sexually reproduced plants under the umbrella of intellectual property protection.

When the Supreme Court deliberated on the Chakrabarty patent, the justices examined a 1952 recodification of the patent statutes. A congressional report that accompanied the legislation (HR 7794) contained a sentence that interpreted section 101 of the patent act: "A person may have 'invented' a machine or a manufacture, which may include anything under the sun that is made by man, but it is not necessarily patentable under section 101 unless the conditions of the title are fulfilled."[12] This phrasing of the Senate committee's report persuaded the court's majority of congressional intent to include living "inventions" under section 101 of the patent act. The court invited Congress to restate its intention if the justices had it wrong and to draw lines on patenting life forms. Since the Chakrabarty decision, Congress has never taken a vote on the patenting of living entities.

The four dissenting judges gave primacy to the plant protection acts as evidence that Congress had not intended the patent laws to encompass living things without its expressed consent in legislation. Speaking for the minority, Justice Brennen wrote: "Because Congress thought it had to legislate in order to make agricultural 'human-made inventions' patentable [in the 1930 legislation] and because the legislation Congress enacted is limited [PVPA in 1970 excluded bacteria], it follows that Congress never meant to make items outside the scope of the legislation patentable."[13]

The phrasing "a machine or manufacture, which may include anything under the sun made by man" in the 1952 Senate report on reauthorization of the patent act does not solve the riddle of whether Congress considered modification of a living thing as something made by man. Breeding, for example, represents a human alteration of species that changes genes, but it doesn't qualify as producing patentable material. It does tell us more about the Court's intention to construct the patent law in favor of the emerging biotechnology industry. A patent is supposed to show how a novel and nonobvious object can be created (or discovered) from the design and basic inputs by someone familiar in the art, in this case the art of genetic engineering. By either inserting or deleting a gene from a microorganism, we have no plan for the organism's complete chemical structure. We cannot assemble its nucleotides; we have not created a reproducible system. If scientists were to have created DNA molecules, cells, ribosomes, and chloroplasts—in other words, construct the organism from fundamental units—then it would meet the ordinary meaning of "manufacture." The patent guidelines clearly provide a far more liberal view of "product of manufacture" as a condition of patentability. A bacterium with 5,000 innate genes and one strategically inserted into its genome (a .02 percent change) is by the courts' logic a "product of manufacture." We should also keep in mind that micoorganisms and chemicals have been awarded patents on grounds that they were selected out of nature or purified, not because they were or could be assembled from basic units.

The 1980 Supreme Court ruling created a broad framework in what the Office of Technology Assessment (OTA) called "statutory construction," which gave the USPTO the legal mandate to award patents on life forms and on parts of living things.[14] On April 12, 1988, the U.S. Patent and Trademark Office issued the first patent (patent number 4,736,866) for a living animal—the Harvard oncomouse. The germ cells and the somatic cells of this animal contain an oncogene that increases the probability of the mouse's developing cancerous tumors. Many other patents on animals and human cell lines were issued in subsequent decades. By 2001, patents were issued on human embryonic stem cells as well as embryos that were developed parthegenetically (without sperm) from unfertilized mammalian eggs and that could be used for cloning animals or for harvesting stem cells. Patent number 6,211,429 was written with language that included all mammalian embryos including humans.[15] While the patent award could cover human embryos, as well as the process of human reproductive cloning, it cannot cover the human product itself because persons are not patentable subject matter. Nevertheless, it is sobering that patent logic has taken society to the point where patenting a human embryo is a realistic outcome.

GENES AS PATENTABLE ENTITIES

Gene patents flow from the logic that they are isolated and modified chemicals. The USPTO finds it irrelevant that genes are the basic chemical constituents of heredity. When a patent is awarded on a gene, is that a patent on an entity or on information? Since genes are ubiquitous and are isolated routinely, the critical piece of information is the location of the DNA sequence and the coding of nucleotides. Once that information becomes available, then those who are familiar with the art can presumably isolate and apply the genetic sequence for some useful function. A patent on the DNA sequence (gene) as a composition of matter gives the patent holder a right to exclude others from using the sequence for any commercial purpose. When a new use is discovered for an already patented DNA sequence, the inventor of the new use may qualify for a patent "notwithstanding that the DNA composition itself is patented."[16] However, the new patent holder may be required to pay a licensing fee to the original patent holder of the DNA sequence.

Notwithstanding the broad consensus within the legal community over the logic of gene patents, the American College of Medical Genetics (ACMG) is one of several professional organizations that has a stated position that "genes and their mutations are naturally occurring substances that should not be patented."[17] The particular concerns over gene patents comprise the following: first, they introduce problematic connections between information and patentable entities; second, they raise the costs of research; third, they restrict the flow of scientific knowledge; and fourth, they create new problems over the interpretation of utility of DNA sequences.

What is the distinction between discovering DNA sequences and products of invention involving those sequences? The courts have ruled that discoveries are not patentable. In 1948, Supreme Court Justice William Douglas, expressing the majority opinion in *Funk Brothers Seed Co. v. Kalo Inoculant Co.,* wrote: "For patents cannot issue for the discovery of the phenomena of nature. . . . The qualities of these bacteria, like the heat of the sun, electricity, or the quality of metals, are part of the storehouse of knowledge of all men. They are manifestations of the laws of nature, free to all men and reserved exclusively to none."[18] Some companies have indicated that they will publish the sequenced segments of the genome. Because decoding the DNA sequences yields information, some people have questioned whether copyrights over the use of the information can be required. If companies did copyright the information from genomic sequencing, then it would not preclude fair use of that information. At most, it may protect the information from being used in published material, but it would not prevent others from developing applications for the sequenced DNA.

The information about the sequenced DNA is one critical piece in developing commercial applications. The nucleotide sequences are patentable when they have utility and are novel. Novelty is lacking when the sequence submitted for a patent has the same chemical structure (e.g., base pairs) already known to the public. Once the exact nucleotide sequences are described, someone familiar with the "prior art" (in this case, molecular genetics) can isolate the sequence and explore applications. More than most research and development fields, applied genomics are information-intensive. While companies cannot patent the information, they can patent the use of the informaton, insofar as it involves the manipulation of actual DNA segments, which are themselves protected by a patent. It is the cDNA that is patented, but it is precisely the form of those DNA segments that is commercially valuable.

A genomics company can profit from sequencing a segment of the genome in two ways. First, they can charge access fees to their sequencing information—if the sequences are not published. Second, they can charge licensing fees for using the biological sequences for commercial applications. Companies that agree to publish the raw sequencing data, however, may still charge fees for access to the programs that interpret the information.

By 2000, the USPTO had issued patents on about 6,000 genes, one-sixth of which were human genes. Once available, genomic information may be used freely by researchers. But the researchers may not freely use the actual genomic material, since that material may be patented. Research activities using a patented product, though technically an infringement, have traditionally been excused from legal liability. Since the early 1800s, the courts have been favorable to experimental-use exemptions when the patented product or process is used "for the sole purpose of gratifying a philosophical taste, or curiosity or for mere amusement."[19] However, in today's world of academic–industry collaborations and the technology transfer interests of universities, genomics companies are less likely to stand passively by while their patented genomic material is used for free. Thus, the tradition of experimental-use exemptions currently does not shelter much human genome research activity, even when conducted in universities and nonprofit institutions.

TRESPASSING ON GENETIC INFORMATION

In 2002, two researchers—Jiangyu Zhu, a Chinese citizen, and his wife, Kayoko Kimbara, a Japanese citizen—were arrested and imprisoned for allegedly stealing genetic material from a Harvard University molecular biology laboratory where they worked in 1999. The couple had accepted postdoctoral positions at Harvard to study the genetic control of the immune system. When the couple left Harvard, they did what researchers often do; namely, they took cell lines with them to their new postdoctoral posts in California—Zhu at University of California, San Diego, and Kimbara at the Scripps Research Institute in La Jolla, California. However, they were investigated by the FBI; charged with intellectual property theft, interstate transportation of stolen property, and conspiracy; and imprisoned in San Diego, pending trial.[20]

The threat of suit has already convinced universities to either pay for licensing of DNA sequences in research activities or to avoid using the sequences altogether. The Myriad Genetics case illustrates the point. Myriad had been awarded patents on the breast cancer genes (BRCA 1 and 2). Mutations in BRCA 1 and BRCA 2 genes have been linked to inherited forms of breast and ovarian cancers. Researchers at the University of Pennsylvania (UP) were studying variations in the BRCA genes as potential markers for identifying individuals susceptible to breast cancer. In 1998, the University of Pennsylvania's genetics laboratory terminated its breast cancer testing program (more than seven hundred tests were conducted a year) when Myriad Genetics threatened them with patent infringement for their use of the BRCA genes.[21]

Myriad's patented gene sequences caused a bit of a stir in the United Kingdom. In February 2000, Myriad licensed the patents on the breast cancer tests to a British biotechnology company, Rosgen. The National Health Service (NHS) was not able to use the knowledge of the gene sequences and their mutations to develop tests without a licensing agreement from Myriad. The alternative was for the NHS to use Rosgen as its testing facility. In 1997, the NHS was charging $960 to screen the two breast cancer genes; in 1998, Myriad Genetics was charging $2,400 for the same test. The higher costs would have made the tests prohibitively expensive for the NHS. Since the gene sequences are in the possession of a private company, national health agencies have to negotiate agreements with the company to avoid patent infringements.

Ironically, two of the BRCA genes codiscoverers—Philip Futreal and Roger Wiseman of the National Institute of Environmental Health Sciences, who had sequenced small fragments of BRCA 1 that were important to Myriad's assemblage of the entire genome—were also charged for use of the sequences in their own research. Agencies like NIH have to negotiate with Myriad for fair licensing fees charged to researchers under their federal grants. After aggressive action taken by NIH in disputing inventorship for BRCA 1, Myriad eventually agreed to include the NIH scientists as coinventors. This enabled the government to receive a 25 percent share of royalties and a role in setting the prices of products derived from the gene sequences.[22]

The story of the BRCA genes is not an isolated case, but it is indicative of how gene patents are having an adverse impact on medical care. For instance, a disease called hemachromatosis, which mainly affects people of northern European descent, causes the body to absorb too much iron from food, eventually resulting in liver damage and heart disorders. Approximately 85 percent of the cases can be traced to two mutations (alleles) in a gene known as HFE. The patent of the gene, first awarded to a California company called Mercator, ended up in another company, which was named Bio-Rad. The patent holder had exclusive rights on testing patients with the disease for the genetic mutations associated with HFE.

In a pilot study run in 1998, a research group found that fewer laboratories were offering the test because the gene was patented. About one-third of the laboratories surveyed stopped offering the genetic tests for HFE because of restrictions imposed on them by the patent holders. The authors of the study wrote: "The patents on HFE had a measurable effect on the development and availability of HFE testing services in the United States as many laboratories that had the capability to perform the test reported not doing so because of the patents."[23] Exclusive licensing by patent holders is viewed as an impediment to the development of better and cheaper tests, but it still leaves open the following question: Even if liberal, nonexclusive licensing were granted to research and nonprofit institutions, why should researchers who do not have commercial interests have to pay licensing fees at all? In other words, why isn't there a broad research exemption for patents?[24]

One legislated research exemption does exist for patent infringement. Congress amended the patent code (PL 98-417) that waived patent infringement for making, using, or selling a patented invention when it was used for the development or submission of information under federal drug laws. Suppose a drug is nearing the end of its twenty-year patent period. A company that is seeking to develop a generic drug based on the patented formulation begins its research on the existing drug prior to the termination period of the patent. According to the law, a company that assembles data to the FDA based on the drug still under patent cannot be cited for patent infringement. The catch is that the testing of the patented material must be *solely* for purposes of meeting FDA requirements.[25]

With the human genome being sequenced by the collective effort of different institutions, each seeking patents on specific segments, the balkanization of the genome leads researchers either to avoid experimenting with certain sequences or to enter "cross-licensing" agreements, which add to the cost of doing research. An editorial in the *Los Angeles Times* wrote: "When patent owners withhold information about human genes from perceived competitors, they prevent the sharing of knowledge that's essential to scientific progress."[26]

In one celebrated case, industrial patent ownership prevented research that could reveal risks of products. Plant ecologist Allison Snow of Ohio State University was funded by the USDA, Pioneer Hi-Bred, and Dow Agrisciences to study transgenic sunflowers. In preliminary results, Snow discovered that the transgene that confers insect resistance to the genetically modified sunflowers can increase the number of seeds produced by wild sunflowers when they are cross-pollinated by the transgenes. The concern to ecologists is that these transgenes could turn wild sunflowers into superweeds. After Snow reported the results of her study to the Ecological Society of America in August 2002, the companies that supported the research refused Snow and her colleagues access to either the transgenes or the seeds from the earlier study—a move that raised the ethical question of whether patent ownership of genes should preempt research on further study of gene flow to wild plants.[27]

Those who oppose gene patents have expressed their sentiments through a variety of evocative metaphors.[28]

- "It's as if somebody just discovered English and allowed the alphabet to be patented."[29]
- "The sequence of my child's genes is not an invention."[30]
- "We certainly have no problem with somebody coming up with a test and patenting it. But they shouldn't be able to patent the knowledge."[31]
- "The notion that some company has a monopoly on my genes is like claiming ownership of the sea."[32]

The decision of the USPTO to patent genes blurs the distinction between invention and knowledge that has guided patent decisions in the past. While it is generally conceded that we cannot patent laws of nature, physical phenomena, and abstract ideas, the sequences of the genome of animals and humans is a discovery about the structure of physical phenomena. The patent law permits the patenting of discoveries from nature, such as penicillin or other naturally produced antibiotics, if they are isolated and purified by humans. The use of a genomic discovery (i.e., the nucleotide sequence of a gene) to study homologous genes or mutations seems less an invention than an application of knowledge, whereas the development of a specific assay from the gene sequences fits the traditional meaning of invention. The upshot of this decision has made every gene sequencer an "inventor" or "discoverer of patentable knowledge," which has inadvertently thrust normal genetic science into entrepreneurship and basic biological knowledge into a realm of intellectual property.

How are these legal and regulatory decisions affecting the attitudes of scientists toward their work? What changes in the norms of scientific practice have resulted from the new paths toward aggressive entrepreneurship taken by universities and their faculty? The next chapter examines the new ethos of scientists in an age of multivested science.

NOTES

1. Eliot Marshall, "Patent on HIV Receptor Provokes an Outcry," *Science* 287 (February 25, 2000): 1375.

2. Paul Jacobs and Peter G. Gosselin, "Experts Fret over the Effect of Gene Patents on Research," *Los Angeles Times*, February 28, 2000.

3. James Madison, letter from Madison to Jefferson (October 17, 1788), in "Federalist No. 43," in *The Federalist Papers*, ed. Clinton Rossiter (New York: Mentor, 1961), 271–272.

4. Madison, "Federalist No. 43," 271–272.

5. Louis M. Guenin, "Norms for Patents Concerning Human and Other Life Forms," *Theoretical Medicine* 17 (1996): 279–314.

6. U.S. Congress, Office of Technology Assessment, *New Developments in Biotechnology: 5. Patenting Life* (Washington, D.C.: USGPO, April, 1989) OTA-BA-370, 51.

7. Norman H. Carey, "Why Genes Can Be Patented," *Nature* 379 (February 8, 1996): 484.

8. Carey, "Why Genes Can Be Patented," 484.

9. Daniel J. Kevles and Leroy Hood, *The Code of Codes* (Cambridge, Mass.: Harvard University Press, 1992), 313.

10. *Diamond v. Chakrabarty*, 477 U.S. 303 (1980).

11. Office of Technology Assessment, *New Developments*, 53.

12. U.S. Congress, Senate, *Revision of Title 35, United States Code,* 82nd Cong., 2nd sess., June 27, 1952, report 1979.

13. *Diamond v. Chakrabarty*, 477 U.S. 303 (1980).

14. Office of Technology Assessment, *New Developments*, 53.

15. Andrew Pollack, "Debate on Human Cloning Turns to Patents," *New York Times,* May 15, 2002. See also, United States Patent, 6,211,429, Machaty and Prather, April 3, 2001. Complete oocyte activation using an oocyte-modifying agent and a reducing agent.

16. Warren A. Kaplan and Sheldon Krimsky, "Patentability of Biotechnology Inventions under the PTO Utility Guidelines: Still Uncertain after All These Years," *Journal of BioLaw & Business*, special supplement: "Intellectual Property" (2001): 24–48.

17. See www.faseb.org/genetics/acmg/pol-34.htm.

18. *Funk Brothers Seed Co. v. Kalo Inoculant Co.*, 333 U.S. 127 (1948).

19. R. Eisenberg, "Patents and the Progress of Science: Exclusive Rights and Experimental Use," *University of Chicago Law Review* 56 (1989): 1017–1086.

20. Rex Dalton, "Scientists Jailed for Alleged Theft from Harvard Laboratory," *Nature* 417 (June 27, 2002): 886.

21. Peter G. Gosselin and Paul Jacobs, "Patent Office Now at Heart of Gene Debate," *Los Angeles Times,* February 7, 2000, A1.

22. Eliot Marshall, "NIH Gets a Share of BRCA 1 Patent," *Science* 267 (February 26, 1995): 1086.

23. Jon F. Merz, Antigone G. Kriss, Debra G. B. Leonard, and Mildred K. Cho, "Diagnostic Testing Fails the Test," *Nature* 415 (February 7, 2002): 577–578.

24. Michael A. Heller and Rebecca S. Eisenberg, "Can Patents Deter Innovation? The Anticommons on Biomedical Research," *Science* 280 (May 1, 1998): 698–701.

25. Office of Technology Assessment, *New Developments*, 59.

26. Editorial, "Gene Discoveries Must Be Shared for the Sake of Society," *Los Angeles Times*, May 14, 2000.

27. Rex Dalton and San Diego, "Superweed Study Falters As Some Firms Deny Access to Transgene," *Nature* 419 (October 17, 2002): 655.

28. Gosselin and Jacobs, "Patent Office Now at Heart," A1.

29. Attributed to Michael S. Watson, American College of Medical Genetics. Quoted in Gosselin and Jacobs, "Patent Office Now at Heart," A1.

30. This quote is attributed to Judith Tsipis, National Tay Sachs and Allied Disease Association, whose son died of Canavan's disease.

31. Fran Visco, president of the National Breast Cancer Coalition.

32. Jonathan King, cited in Andrew Pollack, "Is Everything for Sale?" *New York Times*, June 28, 2000, C1, C12.

5

THE CHANGING ETHOS OF ACADEMIC SCIENCE

More than a half-century ago, the distinguished Columbia University sociologist Robert K. Merton delineated the values inherent in the culture of scientific practice. The normative structure of science, Merton observed, was based on the shared values of free and open exchange of knowledge, the unfettered and disinterested pursuit of truth, and the universal accord among scientists that nature, not culture or religion or economics or politics, was the final arbiter of conflicting views about the physical universe.

At the time that Merton wrote his papers on the values underlying scientific research, between the late 1930s and early 1940s, the legitimacy of science was questioned from two directions. The first direction involved political movements that gained state power, particularly in the Soviet Union and Germany, in which ideological criteria were imposed on the practice of science. In Germany, it meant that Jews were forbidden to hold positions in scientific institutions and that Jewish science (*Juden Wissenschaft*) was not dependable. Meanwhile, in the Soviet Union certain approaches to knowledge and bourgeois scientific theories were dismissed as anticommunist or inconsistent with dialectical materialism—the Marxist-Leninist philosophy of science. The other direction of attack came from social movements that questioned the antihumanitarian fruits of scientific research, especially weapons of mass destruction, labor-displacing technologies, and technologies that placed people under state surveillance and control.

The attacks on science, during a period of global political instability, were an indication of science's rising power. The Enlightenment philosophy still had a powerful hold on many European and American intellectuals who believed in the universality and transcendence of knowledge. But a new breed of sociologists, writing during the turn of the twentieth century, began to explore the cultural and social influences on the production of knowledge. Merton understood both of these traditions. He had been a student of George Sarton, the distinguished historian of science, when he realized that the history of science had no theoretical grounding. During the period in the middle and later 1930s, Merton came upon the idea that science is a social institution. "Once that idea came to me," he said, "I began looking at the normative framework of science."[1] He read scientific diaries, biographies, and a few interviews that were available at the time. In his 1938 paper titled "Science and the Social Order," he began moving toward the concept of institutional scientific norms.[2]

How can knowledge be both a social product and a source of universal truth when so many diverse social structures are involved in the contributions scientists make to knowledge? Merton's solution was to treat science, itself an organized social activity, and society as two demarcated but interdependent social systems. In this respect, we can ask, what are the values inherent in the social system called "science," and how do they relate to the socioeconomic and political systems in which science is embedded? One system's pursuit of universal truths can be distorted by the other system's pursuit of power. This insight led Merton to two questions. What is the normative structure in which science functions? What social system best conforms to the normative structure of science? These questions were not merely intellectual meanderings. Because science is both a beneficiary of social resources and a target of criticism, Merton believed that scientists were "compelled to vindicate the ways of science to man."[3] According to Merton, scientists operate within a social control structure that comprises a reward system (both internal and external to the scientific community)[4] and a set of normative values. He noted that the majority of commentary on his work is directed at the norms with far less emphasis on the reward system of science.

In his investigation into the social structure of science, Merton used the term "ethos" to describe the "complex of values and norms which is held to be binding on the man of science," and it includes prescriptions, proscriptions, preferences, and permissions.[5] He understood that these concepts were not generally codified norms but that they could be inferred from observing the practice and writings of scientists. These norms were all directed at a singular goal—namely, the "extension of certified knowledge."[6] Merton consciously used the term "certified knowledge" in lieu of "truth" since the former implies a social system at work. The community of scientists must agree on the validity of knowledge claims.

For the community of science to function in its most enlightened form, Merton claimed that the ethos of science must include four sets of institutional imperatives, which he defined as universalism, communism (best interpreted as "communalism" or "communitariansm"—not to be confused with the political–economic system), disinterestedness, and organized skepticism. Sociologist John Ziman has explored the culture of science in his book *Real Science*.[7] Referring to Merton's norms, Ziman notes, "These norms are usually presented as traditions rather than moral principles. They are not codified and are not enforced by specific sanctions. They are transmitted in the form of precepts and examples, and eventually incorporated as an ethos into the 'scientific conscience' of individual scientists."[8]

This chapter explores the Mertonian norms of science, and it questions whether they still apply in the aftermath of the biotechnology revolution as academic science became inculcated with corporate values.

MERTONIAN NORMS

The term *universalism* applied to science signifies that certified knowledge transcends the particularity of culture. Objectivity in science demands that truth have no national boundaries. Merton wrote: "The criteria of validity of claims to scientific knowledge are not matters of national taste and culture. Sooner or later, competing claims to validity are settled by the universalistic facts of nature which are consonant with one and not with another theory."[9] To achieve its universalism, science embodies a standardized nomenclature, canonical methods, and an open archive of its achievements.

Geographical boundaries dividing scientists have increasingly become less of an obstacle by the frequency of international meetings, professional associations, and international journals. The pervasiveness and speed of Internet communications has made the goal of a homogeneous global culture of science a closer reality. For instance, U.S. and European physicists consult before either group issues a news release about the discovery of a new elementary particle. In addition, large clinical trials for testing a new drug frequently incorporate human subjects from several nations, and international organizations reach consensus on the causes of disease. Collaboration among scientists across national boundaries has become more commonplace. In some cases, the strength of attachment scientists have to their professional societies and disciplinary affinity groups can even exceed their fidelity to a nation state.

The norm of *universalism* is interpreted by Ziman to include the requirement that "contributions to science should not be excluded because of race, nationality, religion, social status or other irrelevant criteria"[10] and that science does not discriminate in employment. He distinguishes between the private and professional lives of scientists and makes no assumption that scientists, by and large, follow the norms outside of their professional work. It is doubtful that bigotry turns on and off like a spigot. Discriminatory hiring and promotion within the sciences is part of the historical record, even during periods of undeterred scientific progress. Nonetheless, the norm of universalism requires that the nomenclature, methods of investigation, and criteria for certifiable knowledge cannot be culture-specific.

Within the ethos of scientific practice, the norm of *communalism* refers to the common ownership of the fruits of scientific investigation. According to Merton, "The substantive findings of science are a product of social collaboration and are assigned to the community . . . [and] property rights in science are whittled down to a bare minimum by the rationale of the scientific ethic."[11] This norm, according to Ziman, "requires that the fruits of academic science should be regarded as 'public knowledge.'"[12]

What is implied in the common ownership of scientific knowledge is that the results of research should be shared; the information should be freely communicated within and across national boundaries;[13] and a responsibility to the integrity of the "common intellectual fruits" should be ensured. So how exactly do we reconcile "common ownership" of the fruits of knowledge with the practice of acquiring intellectual property rights of one's discoveries and the prevalence of trade secrets among academic researchers? The patenting of inventions arising out of scientific discovery was well entrenched by the time Merton developed his norms. Were his norms an idealized version of science, or were they supposed to reflect scientific practice? Patents may restrict scientific communication for a designated period of time; after which, the discovery must be fully disclosed. In this sense, some would say that patents are not inconsistent with the communitarian values of science. Patents are a compromise between trade secrets and immediate disclosure of scientific results.

Some mistakenly believe that the U.S. patent law makes a clear distinction between discovery and invention. Whereas the fruits of scientific discovery are purported to belong to everyone (Einstein's equations can't be patented), the fruits of invention may be claimed exclusively by the inventor. If science is about discovery and if the boundary between discovery and invention is unambiguous, then the apparent conflict between patent ownership and the communitarian norm may be resolvable.

But there is more than a subtle grayness between discovery and invention. For instance, a biomedical investigator who discovers an antibiotic can, with the proper language, make the case for an invention. The distinction breaks down further when we enter the field of genomics. A scientist who sequences a gene can apply for a patent as if the sequenced gene, in the proper chemical form, is transformed into a "product of manufacture." Merton acknowledged a trend within medicine in the 1940s against patenting the fruits of medical discovery. The most notable case illustrating this point was the discovery of the polio vaccine by Jonas Salk in 1954. Neither Salk nor the March of Dimes, which supported his research, would patent or receive royalties for their discovery/invention.[14]

One of the communitarian arguments against patenting scientific discoveries is that a new result almost always builds on past work and on current collaborations. Any single scientific discovery represents the final step in a genealogical process involving the accretion and synthesis of many threads of information and observation. Following his social systems approach to scientific discovery, Merton remarks that "the substantive findings of science are a product of social collaboration and are assigned to the community."[15] What accrues to the scientist who makes the discovery? "The scientist's claim to 'his' intellectual 'property' is limited to that of recognition and esteem. . . ."[16] Such a view is barely recognizable in this epoch of commercialized science where, notwithstanding the collective contributions of scientists, individuals can accrue the economic benefits for their unique role in the total process. However, in theory, there is room for the communalism of knowledge within the framework of intellectual property and the individuation of rewards—that is, if secrecy in science is abhored in the way that nature abhors a vacuum.

The third norm of science is *disinterestedness.* It requires that scientists apply the methods, perform the analysis, and execute the interpretation of results without considerations of personal gain, ideology, or fidelity to any cause other than the pursuit of truth. Of course, this concept is highly idealistic and quite antithetical to the practice of science. Scientists are not neutral to the outcome of a study in which they may have much at stake. Scientists pose conjectures. Positive results may be publishable, whereas negative outcomes are generally not. Such a scenario means that scientists are not disinterested in the outcome and that they would prefer that the results support their conjecture. But the culture of science demands that results are replicable. This requirement blunts any tendency among most scientists to fudge, since one's career may be dishonored by such behavior.

Although scientists are not disinterested in the outcome of their investigations, they must behave as if they *were* disinterested. They must not allow their bias to affect how they approach their experimental inquiry and how they interpret the results. Not all scientific activity involves original data. Scientists perform reviews of the literature, write commentaries on different subjects, develop theoretical arguments that cite but do not introduce original data, and peer review the grants and papers of other scientists. The norm of "disinterestedness" does not mean that their intellectual interests do not play a role. It is expected that scientists have and exhibit an intellectual standpoint that shapes their perspective. The theories and explanations to which scientists are sympathetic define their intellectual affinities, which are generally available in published papers.

Some observers are resigned to the fact that the norm of disinterestedness in science has all but disappeared. Ziman notes that "post-industrial research has no place for disinterested practices, and post-modern thought has no place for objective ideas."[17] But he muses that science can emerge intact, even without the protection of this norm. Other norms are more critical for the protection of "objective knowledge," he argues. Despite the individual interests of financial arrangements among scientists, the norms that define the production of knowledge as a community-driven process protects knowledge against bias. "The production of objective knowledge then depends less on genuine personal 'disinterestedness' than on the effective operations of other norms, especially the norms of communalism, universalism and scepticism. So long as post-academic science abides by these norms, its long-term cognitive objectivity is not in serious doubt."[18] Merton himself is not convinced that the norm of "disinterestedness" has disappeared. It has surely eroded, he noted, but "these norms are not embedded in granite."[19]

The effects on science and society of the loss of disinterestedness should not be dismissed so easily. First, I agree with Ziman that with sufficient checks and balances within science, convergence to the truth will eventually take place. However, science driven by private interests is laden with bias (see chapter 9). Therefore, it will take more time with repeated replication of results and critical reviews to reach the truth.

A second point is that a culture of science driven by private interests will result in scientists' pursuing knowledge in certain fields for selected problems where there are commercial interests. This trend will come at the expense of other problems that may be of little commercial but of great public interest. In fact, this is already happening in certain fields of biology. For example, significantly more research is done on the uses of chemical pesticides as opposed to biological pest controls.[20] Likewise, vastly more resources are put into the cellular and genetic basis of cancer than into environmental factors.[21]

Third, the loss of *disinterestedness* accompanies the decline in the public orientation of scientists. That is, if the interests of universities become more closely aligned to that of industrial organizations, then so will the interests of academic scientists. (This theme is explored in chapter 11.)

Finally, and this Ziman recognizes, the disappearance of disinterestedness can erode the public's trust in science, regardless of whether "objective knowledge" is affected. But with the decline of public confidence in science, sound public policy is threatened by social disillusionment of any hope for objective knowledge.

The fourth norm in Merton's tetrad is called *organized skepticism*. He says of this norm that one must practice the suspension of judgment "until the facts are at hand" and that the "detached scrutiny of beliefs in terms of empirical and logical criteria" come into play.[22] Skepticism is a virtue. Science must never be subject to a single authority. Merton noted that "most institutions demand unqualified faith, but the institutions of science make skepticism a virtue."[23]

But what is the significance of the adjective "organized"? Once again, Merton understood science as a social system. The spirit of questioning that is so central to the ethos of science places it in direct conflict with dogmatic views of the natural and social worlds, which have not been subject to critical examination. Organized skepticism is not idiosyncratic but is bound by norms of empirical inquiry. Truth is not the default state of authoritative claims, but the result of applying socially established rules of inquiry to a claim that stands up to critical scrutiny and is subject to reexamination.

CHANGING RESEARCH AND DEVELOPMENT FUNDING PATTERNS IN ACADEMIA

How do the Mertonian norms measure up at a time when there appears to be changes in the goals, practices, and social organization of science? The term "academic–industrial complex" describes the new social relations between corporate institutions and scientific researchers in academia.[24] A 1982 report in the journal *Science* wrote: "Scientists who 10 years ago would have snubbed their academic noses at industrial money now eagerly seek it out. . . . The present concentration of industrial interest in academic science is generating no small measure of concern about whether the academy is selling its soul." Much of this change in scientific ethos was precipitated by the aggressive commercial interests in applied molecular genetics and clinical medicine, where the path from discovery to application was rapid and the benefits lucrative.[25]

The incentives put into place by the science policy legislation of the 1980s began to produce effects in the 1990s. Data compiled by the National Science Foundation show that leading research universities are obtaining more of their research and development (R&D) funds from industry.[26]

In 1980, universities and colleges spent $6.5 billion (current dollars) on their R&D. That figure rose to $30.2 billion in 2000, a 467 percent change. In the same period, industry contributions to the R&D budget of academic institutions rose 875 percent from $0.26 to $2.3 billion (constant dollars). During the past three decades, industry R&D support in universities has grown more rapidly than support from all other sources.

The overall contribution of industry-funded R&D in universities rose from 4.1 percent in 1980 (the year the biotechnology industry got underway) to 7.7 percent in 2000. While still a small percentage of the academic R&D budget, 7.7 percent was the highest industry contribution since 1958.[27] The averages for all universities and colleges do not, however, show us the radical changes in funding patterns occurring at several high-profile institutions.

Among the top-ten industry-funded institutions, all but one saw its industry funding grow faster than its total R&D budget. In more than half the institutions, the influence of industry funding was dramatic. For example, Duke University leads the nation in the amount of private sector money it receives. During the 1990s, private funding at Duke grew 280 percent compared to its total R&D, which grew at 85 percent. Other top-ten industry-funded institutions that showed phenomenal growth in attracting more private funding were Georgia Institute of Technology (164 percent) Ohio State University (272 percent), University of Washington (102 percent), University of Texas at Austin (725 percent), and University of California at San Francisco (491 percent). By comparison, the changes in total R&D at these institutions increased between 13 and 60 percent.

The rise of private-funded science in the United Kingdom has been a matter of concern because of the secrecy surrounding commercial research. According to a 2002 report in the *New Scientist*, "Two-thirds of research is now financed by companies. And much of this 'privatized' science is falling into the hands of ever fewer—and ever bigger—global corporations."[28]

The massive infusion of private R&D is changing the character of some institutions. They begin to take on the appearance of contract-research companies that form themselves around major funding troughs of R&D-rich centers around the country. By 2000, Duke University had 31 percent of its budget funded by industry, whereas Georgia Institute of Technology, MIT, Ohio State, Penn State, and Carnegie Mellon had 21, 20, 16, 15 and 15 percent (respectively) of their R&D budgets coming from private sources. As a point of contrast, Harvard University, which has some of the toughest conflict-of-interest guidelines, had a 1.2 percent growth in its industry R&D in the 1990s while its total R&D rose by 29 percent during that time period. In 2000, Harvard's industry funding accounted for 3.6 percent of its total R&D budget.

In response to the massive infusion of private R&D to some highly visible campuses, others are beginning to learn what they need to do to attract this kind of funding. For some, it might mean developing part of its land for a university–industry park. For others, it may mean lowering the ethical bar in reviewing contracts or eliminating impediments to faculty to work on research projects in which they have personal equity. Several smaller research universities have become more highly dependent on industry R&D support than the large research centers. For Alfred University, University of Tulsa, Eastern Virginia Medical School, and Lehigh University, the percentage of their research support coming from industry in 2000 was 48, 32, 24, and 22, respectively.

Industrial liaison programs, modeled on MIT's successful experiment, became popular. Universities introduced technology-transfer offices to monitor, patent, and license potentially profitable discoveries of faculty. The number of patents taken out at major research universities is testimony to the rapid response to federal incentives. Among the top one hundred research universities, the total number of patents awarded annually went from 96 in 1965; 177 in 1974; 408 in 1984; 1,486 in 1994; to 3,200 in 2000. The Association of University Technology Managers published a survey of its own members on patents and licensing. The results indicated that the University of California system, in addition to nine other universities, amassed nearly twelve hundred patents in 2000.[29] Large research universities began to view themselves as incubators for new companies in which they would have an equity ownership. Carnegie Mellon University launched Lycos, an Internet search engine. Boston University invested over $80 million in a biotech company started by its faculty. Duke University was awarded a patent for the gene linked to Alzheimer's disease. Every scientific investigator was a potential entrepreneur. Every gene sequencer had patentable material by decoding a piece of the human genome.

How do these conditions change the mores of science? Are scientists declaring more of their research results as trade secrets? How often do scientists delay publication or withhold data, and what are the reasons behind it? Has commercialization infected science with the idea that the production of knowledge is no longer a commons but a market and that the "market of ideas" is taken literally as "an economic marketplace for research." Twenty years of research are beginning to demonstrate that changes are taking place in scientific behavior and norms. One of the most significant of these is the rejection of "communitarianism" in favor of a system that reinforces limited secrecy and the private appropriation of knowledge. During the period that university-industry partnerships in biotechnology were in their embryonic stages in the early 1980s, science policy analyst David Dickson wrote: "Industry's desire to control the products of university research laboratories—and its need to develop mechanisms to allow the control to be exercised—presents a direct challenge to democratic traditions on which the academic community has traditionally prided itself."[30] Among the losses to the university was the free and open exchange of knowledge.

TRADE SECRECY IN ACADEMIA

A series of surveys undertaken by David Blumenthal and colleagues at Harvard's Center for Health Policy Management provide insight into the changes taking place in the biomedical sciences. In 1986, Blumenthal published the results of a survey of 1,238 faculty members at forty universities that receive the largest amount of federal research funds in the United States. The industry support for biotechnology research in the sample they used amounted to about one-fifth of all available funds. They found that biotechnology faculty with industry support were four times as likely as other biotechnology faculty "to report that trade secrets had resulted from their university research," where "trade secrets" were defined as "information kept secret to protect its proprietary value."[31] The more industrial support of academic research, the greater the number of trade secrets. Moreover, 44 percent of those faculty with industry support and 68 percent of those without such support responded that "to some or to great extent" industry research support undermines intellectual exchange and cooperative activities within departments; in addition, 44 percent of those faculty with industry support and 54 percent of faculty with no industry support reported that such support created unreasonable delays in the publication of new findings.[32]

In a parallel study, Blumenthal and colleagues interviewed 106 biotechnology companies. They found that 41 percent of the biotechnology companies with a university research relationship derived at least one trade secret from the work they supported. The investigators also found that trade secrets were more frequently reported by small firms than they were by Fortune 500 companies.[33] The authors wrote: "A significant minority of firms report the existence of arrangements and behaviors that may threaten traditional university values, such as openness of communication and the unhampered pursuit of knowledge."[34]

A survey of 210 senior executives of life-science companies, taken a decade later between May and September 1994, revealed the following: 92 percent of the firms had some relationship with academic institutions; 82 percent of the firms that fund university scientists require researchers to keep information confidential a period of two to three months or longer, while 47 percent of the respondent firms reported that their contracts with universities protect confidential results for longer than three months to enable the company to file for patents. Also, 88 percent of the respondent executives reported that the confidentiality requirements apply to students.[35] This extended period for sequestering scientific information is considered by some public agencies and professional associations to be excessive and harmful to the progress of science.

Secrecy was found to be more common in industrially supported research compared to research funded by other sources. By comparison with data from a study a decade earlier, the investigators learned that life-science companies were more likely to support academic research in 1994 than in 1984 (57 percent versus 46 percent).

In a study of university–industry research centers carried out at Carnegie Mellon University, half of the centers reported that industry participants could force a delay in publication, and more than a third indicated that industry could have information deleted from papers prior to publication.[36]

The Harvard Health Policy Research and Development Unit published survey results in 1997 that focused exclusively on the degree to which scientists delayed publication of their research results and refused to share results with their colleagues.[37] The team mailed surveys between October 1994 to April 1995 to 3,394 life-science faculty and received responses from 2,167 (64 percent response rate). This survey is the first of its kind to test empirically the degree to which scientists conform to traditional scientific norms. Among the results consistent with reports from other surveys was the finding that almost 20 percent of the faculty reported that publication of their research results was delayed more than six months at least once in the last three years. Nearly 9 percent of the faculty reported that they refused to share research results with other university scientists. Faculty who engage in commercialization were almost three times as likely (31 percent versus 11 percent) to report publication delays longer than six months. They were also more likely to have denied other scientists access to research results or biomaterials compared to those who were not engaged in commercial activities. The authors of the study commented: "The fact that 34 percent of faculty have been denied access to research results suggests that data withholding has affected many life-science faculty."[38]

In a more recent study, a team of investigators led by Eric Campbell of the Harvard Medical School undertook a mail survey of nearly 3,000 scientists and received 1,849 responses, yielding a 64 percent response rate.[39] They asked scientists whether at least one of their requests for information, data, or materials related to published research results was denied in the past three years. Of the geneticists that responded, 47 percent did so in the affirmative. Because they were denied information, 28 percent of the geneticists reported that they were unable to confirm the accuracy of published results.

Twelve percent of the respondents indicated that they denied another scientist's request at least once in that three-year period. This disparity between the number of those respondents who stated that they were denied data against those who deny others data is curious. Either small numbers of scientists with highly prized data are hoarding their information in response to the many scientists who make requests, or the respondents may have forgotten or undercounted their own denials of data requests.

Of the 1,800 scientists who responded to the survey, 31 percent indicated that they were engaged in commercial activities, and as many as 27 percent of the respondents cited commercial reasons for why they withheld data or biological materials from other scientists who made a request. Over a third (35 percent) of geneticists in the survey believe that data withholding is increasing in their community.

Taken together, these studies provide convincing evidence that commercialization has taken a foothold in the university. At least within the life sciences, these studies demonstrate greater secrecy among colleagues, a significant failure of scientific exchange in the community, and a pattern of delayed publication. These results suggest a new tolerance among academic scientists for commercializing first and sharing later.

Another consequence of commercial partnerships with universities is the interest that investors in academic research have shown for creating private enterprise zones within the nonprofit knowledge sectors. The norm of communalism—that "substantial findings of science are a product of social collaboration and are assigned to the community"[40]—has become secondary to the norm of private appropriation of one's intellectual labor.

As noted in chapter 4, the patent system has blurred the distinction between discovery and invention, and this is especially prevalent in biological research. On January 5, 2001, the U.S. Patent and Trademark Office (USPTO) published guidelines for assigning patents to genomic sequences.[41] Some commentators wrote the USPTO that they believe inventions, not discoveries, were patentable subject matter, and that since newly found gene sequences were discoveries and not inventions, they should not be patentable. The USPTO's response was that patent law (35 U.S.C. 101) explicitly includes discovery and invention: "Whoever invents or discovers any new and useful process, machine, manufacture, or composition of matter, or any new and improvement thereof, may obtain a patent, subject to the conditions and requirements of this title." The knowledge obtained from discovered gene sequences, so long as those sequences are isolated from their natural state, is patentable. Once patented, those sequences are not the shared intellectual property of the scientific community. Industrial as well as academic scientists may have to negotiate a licensing agreement to use the sequences, even if the use is in pure research. While many scientists still find it incredulous, as noted in chapter 4, no statutory research exemption from patent infringement exists for scientific experiments using intellectual property.[42]

The issue of the privatization of knowledge has been especially contentious with respect to the Human Genome Project. This multibillion-dollar effort to sequence the thirty to fifty thousand or so genes in the human genome[43] has brought out different views about whether the information should be in the public domain. The National Institutes of Health had initially sought patents for public-funded genomic information so as to keep the information in the public domain. Subsequently, NIH withdrew its patent applications, giving the market system free reign on the proprietary nature of genomic information. It is a further example where the federal government has abandoned the "knowledge commons." Erosion of such a concept has spilled over to the scientific journals.

Publication in science has generally operated under the principle that primary data that supports a scientific result should be open to members of the scientific community. That principle was renegotiated under an unorthodox agreement between the editors of the journal *Science* and the firm Celera Genomics of Rockville, Maryland. Scientists at Celera contributed an article to *Science* on the human genomic code, which was accepted for publication. The policy of the journal is to ensure that the supporting data of published results are freely available to the scientific community. For genomic data, this rule has meant depositing the data on a publicly accessible database. In this case, the editors of the journal struck a balance between the norms of the scientific community for free and open exchange of knowledge and the company's proprietary interests. Under the agreement, the company was allowed to hold the supporting data for the article in its own systems and to make the data available to other scientists on the condition that they meet certain requirements.[44]

The ownership of knowledge took another unusual turn when an independent California mathematician received the first patent on a mathematical result. The mathematician was awarded his patent for both the technique of finding certain kinds of prime numbers (numbers that are only divisible evenly by one and itself) and for the use of two of those prime numbers, but only when they are used together. Since the numbers are 150 and 300 digits long, respectively, it is unlikely that there would be much patent infringement. The patent gives its owner the legal right to sue for infringement by anyone who uses the numbers without permission.[45]

POSTACADEMIC SCIENCE

The British sociologist John Ziman distinguishes between academic and industrial science. They differ, according to Ziman, in their social goals and social organization: "Academic science can be described in terms of the Mertonian norms. . . . Industrial science contravenes the norms at almost every point."[46] The acronym that Ziman uses to characterize industrial science is PLACE, which stands for Proprietary, Local, Authoritarian, Commercial, and Expert. Industrial science produces proprietary knowledge, usually commissioned; addresses local technical problems, rather than broad foundational problems; operates under strict managerial authority to achieve practical goals; and employs expert problem solvers, rather than people who pursue their own self-directed investigations.

The term postacademic science is used by Ziman to describe the state under which industrial science mixes with academic science. "It implies the establishment within academic science of a number of practices that are essentially foreign to its culture."[47] Once the hybridization of science takes form, postacademic scientists perform dual roles. Ziman comments rather flippantly, "On Mondays, Wednesdays and Fridays, so to speak, they are producing public knowledge under traditional academic rules, on Tuesdays and Thursdays they are employed to produce private knowledge under commercial conditions."[48]

In actuality, the transition in roles between academic science and postacademic science is far more seemless than Ziman describes. More typically, scientists receive funding from companies (in which they may have equity or serve as principles) to carry out research in their university laboratories. To the graduate student or postdoc working in the lab, it may not even be obvious what activities are commercially funded and what are publicly funded. The rules of the laboratory are the same. Trade secrecy operates for both commercially funded and publicly funded research since either may yield proprietary knowledge.

Ziman's confidence in the protection of "objective knowledge" within the framework of postacademic science comes from the primacy of norms like universalism and skepticism. The loss of "disinterestedness" brings with it some decline in trust both from professional circles and the public, but he assures us that "objective knowledge" can prevail in a culture infused with multivested scientists pursuing truth and profits.

The concept of postacademic science suggests that university science has evolved from its pure form to a hybridized form and that this newly evolved form of science carries with it a new ethos. Ziman's conception of science fails to consider two things. First, not all the sciences have evolved into this state. Some natural and social sciences have not become tethered to commercial interests and have therefore not shaken off the norm of "disinterestedness." Second, various sectors within science go through different cycles depending upon the external incentives. Military research may be strong during certain periods and weak during others. The passage of the Homeland Security Act in 2002 will once again draw universities into the defense mission because of the massive infusion of new funds. The American university has never been a pure embodiment of the ideal of what I have called the "Classical Personality." Multiple personalities always coexist in the same institution. External incentives may activate one or more of the personalities and elevate the norms associated with those at different times.

I believe all the elements of the postacademic university have always existed in some nascent form in academia. The classical and industrial ideals, however, continue to coexist. Some disciplines at some universities have had a longer tradition of entrepreneurial activities, including patent development and collaboration with industrial sectors. For other disciplines and other universities, the realization of the "Baconian Ideal," which exists at some preemergent level, may be novel, thus displacing the "Classical Ideal" at least for the time being. We should not assume, though, that some natural development occurs from academic to postacademic science. The corporate influence on scientists is neither natural nor inevitable, and the chapters ahead will explore the societal implications of the disappearance of disinterestedness in science.

One of the most critical roles that university scientists play in modern life, beyond their contributions to basic knowledge, is in policy and law. Our lawmakers, regulators, judges, and juries depend on scientists to provide dispassionate analyses on the state of our knowledge in areas germane to the passage of policy and to tort litigation. Toward this end, an elaborate system of federal advisory committees provides the input into countless policy decisions. The next chapter looks at how conflicts of interest compromise the integrity of the federal advisory committee process.

NOTES

1. Robert K. Merton, personal communication, October 7, 2002.
2. Robert K. Merton, "Science and the Social Order," *Philosophy of Science* 5 (1938): 321–337.
3. Robert K. Merton, *Social Theory and Social Structure* (New York: The Free Press, 1968), 604.
4. Robert K. Merton, "Priorities in Scientific Discovery: A Chapter in the Sociology of Science," *American Sociological Review* 21 (December 1957): 635–659.
5. Merton, "Priorities in Scientific Discovery," 605.
6. Merton, "Priorities in Scientific Discovery," 606.
7. John Ziman, *Real Science* (Cambridge, U.K.: Cambridge University Press, 2000).
8. Ziman, *Real Science*, 31.
9. Merton, *Social Theory and Social Structure*, 608.
10. Ziman, *Real Science*, 36.
11. Merton, *Social Theory and Social Structure*, 610.

12. Ziman, *Real Science*, 33.

13. "Secrecy is the antithesis of this norm; full and open communication its enactment." Merton, *Social Theory and Social Structure*, 611.

14. Seth Shulman, *Owning the Future* (Boston: Houghton Mifflen Co., 1999), 54.

15. Merton, *Social Theory and Social Structure*, 610.

16. Merton, *Social Theory and Social Structure*, 610.

17. Ziman, *Real Science*, 180.

18. Ziman, *Real Science*, 174.

19. Robert K. Merton, personal communication, October 7, 2002.

20. In our study of the classification of articles in the journal *Weed Science* between 1983–1993, we found research articles on herbicides vastly outnumbering the articles on nonchemical weed control. Sheldon Krimsky and Roger Wrubel, *Agricultural Biotechnology and the Environment* (Urbana: University of Illinois Press, 1996), 52–53.

21. Samuel Epstein, a professor of occupational and environmental medicine at the University of Illinois, determined that in 1998 the American Cancer Society allocated "well under one hundreth of one percent of the Society's total budget on environmental carcinogenesis." www.nutrition4health.org/NOHAnews/NNSO1_EpsteinCancer.htm (accessed May 23, 2002).

22. Merton, *Social Theory and Social Structure*, 614.

23. Merton, *Social Theory and Social Structure*, 601.

24. Martin Kenney coined the term "university–industrial complex" in the title of his book, *Biotechnology: The University–Industrial Complex* (New Haven: Yale University Press, 1986).

25. Barbara J. Culliton, "The Academic–Industrial Complex," *Science* 216 (May 28, 1982): 960.

26. National Science Foundation, Directorate for Social, Behavioral and Economic Science, "How Has the Field Mix of Academic R&D Changed?" (December 2, 1999), 99-309. See table B-32, "Total R&D Expenditures at Universities and Colleges: Fiscal Years 1992–1999"; Table B-38, "Industry-Sponsored R&D Expenditures at Universities and Colleges: Fiscal Years 1992–1999." Accessed at www.nsf.gov/search97cgi/vtopic.

27. National Science Foundation, *Science and Engineering Indicators, 2002* (Washington, D.C.: NSF, 2002), appendix table 5-2, "Support for Academic R&D, by Sector: 1953–2000."

28. David Concor, "Corporate Science versus the Right to Know," *New Scientist* 173 (March 16, 2002): 14.

29. The Association of University Technology Managers, *AUTM Licensing Survey: FY2000* (2002), 115–119, at www.uutm.net/survey/2000/summary.noe.pdf (accessed on September 15, 2002).

30. David Dickson, *The New Politics of Science* (New York: Pantheon, 1984), 105–106.

31. David Blumenthal, Michael Gluck, Karen S. Louis, Michael A. Soto, and David Wise, "University–Industry Research Relationships in Biotechnology: Implications for the University," *Science* 232 (June 13, 1986): 1361–1366.

32. Blumenthal et al., "University–Industry Research Relationships," 1365.

33. David Blumenthal, Michael Gluck, Karen S. Louis, and David Wise, "Industrial Support of University Research in Biotechnology," *Science* 231 (January 17, 1986): 242–246.

34. Blumenthal et al., "Industrial Support of University Research," 246.

35. David Blumenthal, Nancyanne Causine, Eric Campbell et al., "Relationships between Academic Institutions and Industry in the Life Science: An Industry Survey," *New England Journal of Medicine* 334 (February 8, 1996): 368–373.

36. Richard Florida, "The Role of the University: Leveraging Talent, Not Technology," *Issues in Science and Technology* 15 (Summer 1999): 67–73.

37. David Blumenthal, Eric G. Campbell, Melissa S. Anderson, Nancyanne Causino, "Withholding Research Results in Academic Life Science: Evidence from a National Survey of Faculty," *Journal of the American Medical Association* 277 (April 16, 1997): 1224–1228.

38. Blumenthal et al., "Withholding Research Results," 1224–1228.

39. Eric G. Campbell, Brian R. Clarridge, Manjusha Gokhale et al., "Data Withholding in Academic Genetics," *Journal of the American Medical Association* 287 (January 23/30, 2002): 473–480.

40. Robert K. Merton, "A Note on Science and Democracy," *Journal of Legal and Political Sociology* 1 (1942): 115–126.

41. U.S. Patent and Trademark Office, Department of Commerce, "Utility Examination Guidelines," *Federal Register* 66 (January 5, 2001): 1092–1099.

42. M. Patricia Thayer and Richard A. DeLiberty, "The Research Exemption to Patent Infringement: The Time Has Come for Legislation," *The Journal of Biolaw and Business* 4, no. 1 (2000): 15–22.

43. The estimate of the number of human genes has been reduced from its initial figure of 100,000 to as low as 30,000.

44. Peter G. Gosselin, "Deal on Publishing Genome Data Criticized," *Los Angeles Times*, December 8, 2000, A7.

45. Simson Garfunkle, "A Prime Argument in Patent Debate," *Boston Globe,* April 6, 1995, A69–70.

46. Ziman, *Real Science*, 78.

47. Ziman, *Real Science*, 79.

48. Ziman, *Real Science*, 116.

6

THE REDEMPTION OF FEDERAL ADVISORY COMMITTEES

During the late 1970s, I was appointed to an advisory committee of the National Institutes of Health (NIH) by Joseph Califano, who was then Secretary of Health, Education, and Welfare (which was later changed to Health and Human Services). The committee was given the rather arcane name Recombinant DNA Molecule Advisory Committee, or the "RAC" as it was more popularly called. The charge of the committee was to advise the director of the NIH on setting guidelines for government-funded scientists who were using the new gene splicing (recombinant DNA) techniques. During the 1970s, when the advisory committee was first established, and throughout the 1980s, gene-splicing experiments were hotly debated. The function of the RAC was to assess and reassess the potential risks of certain classes of experiments and to recommend revisions of the guidelines first issued by NIH in 1976.

Like scores of other federal advisory committees, candidates for appointment to RAC had to file disclosure forms with the NIH, listing any relationships, financial or otherwise, that might be viewed as a potential conflict of interest. While I submitted my form, I had no idea what, if any, conflicts of interest were possessed by my colleagues. This committee of twenty-five members met several times a year and debated the risks of gene-splicing experiments, discussed the release of genetically modified organisms into the environment, and reviewed confidential business documents of companies who participated in NIH's voluntary compliance program. Under the program, companies sought the imprimatur of RAC for genetic-engineering experiments they wished to undertake.

After serving on the committee for two years, I learned that several of the scientists on the RAC had financial interests in biotechnology companies.[1] This information was not available to the media, which had covered many aspects of the committee's decisions. It was also hidden from the public members of the committee who had been selected to create the impression that this group was not a self-serving committee of insiders. Even a formal freedom of information (FOI) request could not bring conflict-of-information (COI) information out into the sunshine. The only reason that the information became available was that some committee members were principals in companies that issued a public stock offering. Their names appeared in the media.

THE PUBLIC ROLE OF SCIENCE

How does the scientific advisory committee process function in the United States? What safeguards are in place to prevent expert panels from serving private interests rather than public interests? What is the role of disclosure, and is it effective? This chapter explores the current state of conflict-of-interest policies on scientific advisory committees of the federal government.

The system of decision making in the United States depends heavily on the use of scientific experts, many from independent nonprofit institutions such as universities and research institutes. The public assumes, sometimes naively, that when academic scientists participate on these committees, they do not bring a special interest to the decision process, only their expertise. Agencies like the Food and Drug Administration, the U.S. Department of Agriculture, the Centers for Disease Control, the Department of Energy, the Federal Aviation Administration, and the Environmental Protection Agency are in the business of promulgating regulations. Other agencies, like the National Institutes of Health and the National Science Foundation, are in the business of awarding grants to scientists. They set up advisory panels to peer review proposals or to set research agendas.

Hundreds of advisory panels exist throughout government, some short-term, some long-term. The idea behind the system of advisory committees is sound. Government scientists and bureaucrats cannot possibly assemble the expert knowledge necessary to address the diversity of technical problems under the government's responsibility. The use of advisory committees not only taps the best available knowledge but also brings diversity into the federal decision-making process. Moreover, as one government report noted, "Advisory committees continue to represent part of federal efforts to increase public participation."[2]

In fiscal year 1998 (between October 1997–1998), fifty-five federal departments and agencies sponsored 939 advisory committees. Of those, fifty advisory committees, including twenty-one established by the Congress, directly advised the President. A total of 41,259 individuals served as committee members, among whom one-third (14,860) were appointed by the Department of Health and Human Services. In sum, 5,852 meetings were held, and 973 reports were issued. During that fiscal year, $180.6 million was expended to fund the costs of the advisory committees, including compensation to committee members, travel reimbursement, per diem expenses, staff support, administrative overhead, and consulting fees. About half the costs went to administrative overhead to run the committees.[3] In the year 2000, the FDA had thirty-two standing advisory committees while the NIH maintained over 140 advisory committees.

LAWS GOVERNING FEDERAL ADVISORY COMMITTEES

The primary federal legislation governing advisory committees, task forces, and councils—established by either the president, the federal agencies, or Congress—is the Federal Advisory Committee Act (FACA), signed into law in 1972. Under FACA, membership on committees must be fairly balanced in terms of points of view and functions of the committee. The statute explicitly forbids FACA committees to be inappropriately influenced by special interests and that provisions be in place "to assure that the advice and recommendations of the advisory committee will not be inappropriately influenced by the appointing authority or by any special interest, but will instead be the result of the advisory committee's independent judgment."[4]

The ethical guidelines that are applied to advisory committees are found in the U.S. Code of Federal Regulations (18 U.S.C., sec. 202–209). Advisory committee members are considered Special Government Employees (SGEs) as long as their service on the committee does not exceed 130 days a year. A consultant on a federal advisory committee is considered an SGE even though the consultant has not taken an oath of office, receives no salary, and does not have permanent employment with the government. As an SGE, scientific advisory committee members must comply with federal conflict-of-interest laws. The purpose of conflict-of-interest statutes is to prohibit government officials from engaging in conduct that would be self-serving and inimical to the general public's best interests. In particular, the statutes and regulations are intended to prevent any employee of the Executive Branch or an independent agency, including federal advisory committee members, from permitting personal interests to affect their decisions and to protect the government proceses from "actual or apparent conflicts of interest."[5]

Executive Order 12838, titled "Termination and Limitation of Federal Advisory Committees," issued by President Clinton in 1993 directed federal agencies to terminate advisory committees whose missions were no longer supportive of national interests or whose work was duplicative of other committees or agencies. The presidential order sought to radically reduce the number of advisory committees used in government and to set stringent restrictions on the creation of new committees. The order stated:

> Executive departments and agencies shall not create or sponsor a new advisory committee subject to FACA unless the committee is required by statute or the agency head (a) finds that compelling considerations necessitate creation of such a committee, and (b) receives the approval of the Director of the Office of Management and Budget. Such approval shall be granted only sparingly and only if compelled by consideration of national security, health or safety, or similar national interests.[6]

In December 1996, the Office of Government Ethics issued regulations that interpreted 18 U.S.C., section 208. Specifically, the regulations provided guidance to agencies in interpreting COI laws, in handling conflicts of interest, and in granting waivers and authorizing exemptions to conflict-of-interest prohibitions.[7]

Amendments to the Federal Advisory Committee Act were passed in December 1997 (PL 105–153) requiring the General Services Administration (GSA) to report to Congress on implementation of the new provisions of the law, which included the role of advisory committees in the National Academy of Sciences. Federal regulations prohibit employees, including SGEs, from taking part in a decision-making process when they have a personal interest in the subject matter under discussion: "A federal employee may not personally and substantially participate in an official capacity in any particular matter, which to his knowledge, he or any other person (whose interests are imputed to the employee under 18 U.S.C., sec. 208) has a financial interest if the particular matter will have a direct and predictable effect on that interest."[8] The catch in this regulation is that employees with conflicts of interest can participate if they receive a waiver.

A waiver can be issued by the appointing officer of an agency, usually the agency head, for two reasons. First, the appointing officer may determine that the financial interest of the prospective advisory committee member is remote, inconsequential, insubstantial, or that it is unlikely to affect the integrity of the services the advisor is expected to deliver. Second, a prospective advisory committee member, who would be disqualified for his or her conflicts of interest, may be allowed to serve if the appointing officer determines that the need for that person's services outweighs the potential conflicts of interest of the advisor. This provision leaves considerable latitude for the appointing officer to waive a conflict of interest.

Prior to 1989, each federal agency had its own interpretation of the statute on waivers (U.S.C., sec. 208[b][2]) for potential advisory committee members who had a financial conflict of interest. That changed somewhat with passage of the Ethics Reform Act of 1989, which reduced the authority of individual agencies to adopt COI exemption criteria by granting that authority to the Office of Government Ethics (OGE). The act stated that the director of the OGE may issue regulations that exempt from the general prohibtion "financial interests which are too remote or too inconsequential to affect the integrity of the services of the employees to which the prohibition applies." In 1996, the OGE issued general guidelines for federal agencies to use when reviewing the COI disqualifications of an individual who is being considered for service on an advisory committee.

The conflicts of interest of committee members are usually not made public. The Ethics in Government Act does not require agencies to disclose publicly an advisory committee member's principal employment, contractual relationships, or investments that may be relevant to the issues discussed. It is only when a controversy breaks out that these conflicts of interest and the matter of the waiver reach public attention. Each of the government agencies interprets its legal responsibility under the government's ethics rules according to the political sensitivity of its own activities. For instance, the Food and Drug Administration (FDA) deals with issues that are continually in the public spotlight. The FDA advisory committees have been the subject of several investigative reports that raise concerns about the integrity of the process.

FDA ADVISORY COMMITTEES

The U.S Food and Drug Administration is situated administratively within the Department of Health and Human Services (DHHS). Under the DHHS regulations (5 C.F.R. 2635), advisory committee members may not participate in matters that are likely to have a "direct and predictable effect" on the financial interests of a person with whom he or she has a relationship (members of household, close friend, or employer). Like other federal agencies, the FDA can exercise the waiver provision of the U.S. ethics statutes on conflicts of interest.

In the fall of 2000, *USA Today* investigative reporters studied eighteen expert advisory committees established by the FDA's Center for Drug Evaluation and Research, which met between January 1, 1998, and June 30, 2000. These committees typically make recommendations to the agency on whether new drugs should be approved and, if so, on what conditions accompany such approval. The journalists were able to get information on the number of committee members with financial interests and the number of waivers issued to committee members. Prior to 1992, the FDA made public the details of advisory members' financial conflicts of interest. After a series of controversies, including one involving Prozac, the FDA stopped disclosing the particulars of committee members' conflicts of interest for the reputed purpose of protecting their privacy. But they are still required to disclose when a committee member has a conflict of interest in the subject of a meeting. *USA Today* found that there were 159 FDA advisory committee meetings held during the eighteen-month period they studied. Approximately 250 committee members appeared 1,620 times during the 159 meetings.[9]

At least one advisory committee member had a financial stake in the topic under review at 146 of the 159 meetings (92 percent). At 88 of the meetings (50 percent), at least half of the advisory committee members had financial interests in the product being evaluated.[10] The financial conflicts of interest were most frequent (92 percent of the members had conflicts) at the fifty-seven meetings at which broader policy issues were discussed; however, at the 102 meetings that dealt with a specific drug application, 33 percent of the experts had a conflict of interest.[11]

USA Today reported that "more than half of the experts hired to advise the government on the safety and effectiveness of medicine have financial relationships with the pharmaceutical companies that will be helped or hurt by their decisions."[12] It noted that 54 percent of the time advisers to the FDA have "a direct financial interest in the drug or topic they are asked to evaluate."[13] Most of the financial conflicts involved stock ownership, consulting fees, and research grants.

How is it that so many members of the eighteen FDA advisory committees who held conflicts of interest were permitted to participate and vote on committee matters? The law states that an advisory committee member has a conflict of interest when an action taken by the committee could have the "direct and predictable effect" of causing the member to have a financial gain or loss. The FDA officials issued waivers when they believe the expert's potential contribution to the committee outweighs the seriousness of the conflict. Of the 1,620 member appearances, 803 (50 percent) had waivers for COIs, 71 (4.4 percent) had disclosures of financial interests not resulting in waivers, and the remaining 746 (46 percent) constituted member appearances who had no COIs. Did over 800 committee members with conflicts of interests justify waivers? Is the FDA issuing so many waivers because the reservoir of independent experts is declining so rapidly that there are fewer experts left without ties to the industry that the FDA regulates?

Whatever the reason for all the waivers, they hardly foster public confidence in the system through which government receives expert advice. Some argue that the very least our government agencies owe to the public is to make the advisory system totally transparent with respect to conflicts of interest. It is generally through litigation and congressional investigation, however, that the conflicts among members of advisory committees are exposed, usually after decisions are made.

TWO FEDERAL ADVISORY COMMITTEES AND A VACCINE

As discussed in chapter 2, a vaccine manufactured by Wyeth Lederle for rotaviruses, one of the leading causes of acute gastroenteritis, was approved and then withdrawn, all within about a year. This case precipitated a federal investigation into the advisory committees that were central to the decision making. The results of the investigation revealed an advisory process that was endemic with conflicts of interest and that raised serious questions about the integrity of the process.

In August 1999, the U.S. House of Representatives Committee on Government Reform began an investigation into federal vaccine policy, which focused on the conflicts of interest of policy makers in this specific case. The committee reviewed financial disclosure forms, scrutinized minutes of meetings, and interviewed advisory committee members. The majority staff report was released in August 2000.[14]

The committee investigation revealed that the conflict-of-interest rules employed by two agencies of government responsible for vaccine policy making, the FDA, and the Centers for Disease Control (CDC), "have been weak, enforcement has been lax, and committee members with substantial ties to pharmaceutical companies have been given waivers to participate in committee proceedings."[15] In a letter to DHHS Secretary Donna Shalala, House committee chair Dan Burton said: "It has become clear over the course of this investigation that the VRBPAC [the FDA's Vaccine and Related Biological Products Advisory Committee] and the ACIP [the CDC's Advisory Committee on Immunization Practices] are dominated by individuals with close working relationships with the vaccine producers."[16]

The Committee on Government Reform found that three out of the five full-time FDA advisory committee members who voted for the vaccine had financial ties to either Wyeth Lederle or two other companies developing competitive rotavirus vaccines. Out of the eight CDC advisory committee members who supported the vaccine, four had financial interests with the same companies. The committee studied the conflicts of interests among VRBPAC members at a critical meeting when the rotavirus vaccine was voted favorably. The committee reported that "the overwhelming majority of members, both voting members and consultants, have substantial ties to the pharmaceutical industry."[17] Table 6.1 shows the conflicts of interests found among participants in the rotavirus decision process.

The CDC granted blanket waivers to all its ACIP members for the entire year, regardless of the nature of the conflict of interest. The report found that "ACIP members are allowed to vote on vaccine recommendations, even when they have financial ties to drug companies developing related or similar vaccines."[18]

The conflicts of interest found among the CDC's vaccine advisory committee included one member who owned 600 shares of stock in Merck, which was valued at $33,800; another who shared the patent on rotavirus vaccine in de-

Table 6.1. FDA Vaccine Advisory Committee Conflicts of Interest[19]

Permanent Committee Members (n = 15)	*Temporary Voting Members (n = 5)*
Person A: Patent holder of Rotashield; excluded from deliberations. COI not waived.	Person G: Vanderbilt University scientist whose university received extensive grants and contracts from drug companies. No COI reported.
Person B: Led discussion on the vaccine; owned about $20,000 stock in Merck, a manufacturer of a competititve rotavirus vaccine. COI waived.	*Consultants (n = 3)*
Person C: Consumer representative; advocate for vaccine; received travel expenses and honoraria from Merck. No waiver needed.	Person H: Received frequent travel reimbursements and honoraria from drug companies including Merck; fundraiser for Johns Hopkins University; sought start-up funds from vaccine manufacturers for a vaccine institute. COI waived.
Person D: University of Rochester scientist received over $9 million from NIAID for vaccine research; NIAID licensed the right to develop Rotashield to Wyeth Corp. No waiver needed.	Person I: Owned about $26,000 of stock in Merck; served on the Merck advisory board. COI waived.
Person E: Received a contract from Wyeth Lederle of over $250,000 per year between 1996–1998 for study of pneumococcal vaccines. No waiver needed.	
Person F: Baylor College of Medicine scientist received funds for development of rotavirus vaccines; paid $75,000 grant from American Home Products. COI waived.	

velopment by Merck, who received a $350,000 grant from Merck, and who consulted with Merck for rotavirus vaccine development; a third person who had a contract with Merck's vaccine division and had received funds from other companies; a fourth who was a professor at Vanderbilt University School of Medicine and served on a Merck committee and whose wife consulted for subsidiaries of American Home Products; a fifth who worked for a company that was participating in vaccine studies for Merck and other companies; and a sixth who received educational grants from Merck and SmithKline Beecham.

Even those members who were disqualified from voting on recommendations because of their financial conflicts of interest were permitted to fully participate in the discussion leading up to a vote. These individuals at times gave their voting preference during deliberations, which is considered to be a breach of ethical guidelines.

The findings of the House committee showed that conflicts of interest are deeply embedded in the vaccine program at FDA and CDC. The report stated: "The FDA standards defining conflicts of interest are ridiculously broad. For example, a committee member who owns $25,000 in stock in an affected company is deemed to have a low involvement interest, which is usually automatically waived. In fact, a member could own up to $100,000 in stock in an affected company, a 'medium involvement' by FDA standards, and that conflict would generally be waived."[20] As a matter of record, a committee member who received $250,000 a year from the maker of a rotavirus vaccine was granted a waiver and actually voted in the VRBPAC deliberations. The House committee concluded that the CDC has virtually no conflict-of-interest standards because it automatically waives all ACIP members.

In 1992, the National Academy of Science's Institute of Medicine recommended that the FDA initiate some changes in its use of expert advisory committees to avoid potential conflicts of interest.[21] Eight years later, the FDA continues to be cited for the egregious conflicts of interests its scientific advisers have with the products they are called upon to evaluate.

The House committee's investigation into the rotavirus affair, in conjunction with the revelations about FDA's use of scientific advisers, begins to paint a portrait of an agency where conflicts of interest have become normalized in the process of drug evaluation. The paradox that underlies the process is as follows: how can there be disinterested evaluators when the leading experts in the field are either working for industry or are academic scientists who are highly paid consultants to companies manufacturing the drugs?

Under a new law passed in 1997, the FDA is allowed to add official industry representatives to advisory committees. They will be allowed to participate in deliberations, but they will not be allowed to vote. This change will further bias FDA panels in favor of drug manufacturers at a time when there is already formidable industry consultant bias represented on the panels.

AMERICA'S DIETARY GUIDELINES

I can still remember those early U.S. Department of Agriculture (USDA) public service commercials on television that outlined seven basic food groups and advised viewers to balance their diets with sufficient meat, fish, vegetables, fruits, milk and cheese, nuts, and grains. That campaign was well before the revolution in junk food. Since the 1950s, the federal government has upscaled its recommendations from dietary guidance to dietary guidelines. The USDA went from issuing nutritional guidance on food groups (to help consumers choose a balanced diet) to promulgating dietary guidelines designed to reduce the risk of major chronic diseases.

The first set of dietary guidelines was issued in 1980. Ten years later, Congress passed the National Nutrition Monitoring and Research Act (PL 101–445), which requires publication of the federal dietary guidelines every five years beginning in 1995.

A pro-vegan organization called the Physicians Committee for Responsible Medicine (PCRM; founded in 1985) was the plaintiff in a suit filed against the USDA on December 15, 1999. The PCRM argued that the agency violated the Federal Advisory Committee Act and the Freedom of Information Act by not revealing the conflicts of interests of the members of its Dietary Guidelines Advisory Committee, which is responsible under federal law for revising the dietary guidelines every five years. The guidelines, which were revised in the summer of 2000, serve as the basis for all federal food assistance and nutrition programs, including the School Lunch Program, the federal Food Stamp Program, the School Breakfast Program, and the Special Supplemental Nutrition Program for Women, Infants, and Children. The PCRM argued that the USDA had intentionally withheld information about the eleven members of its advisory committee, six of whom it alleged currently have or recently had "inappropriate financial ties" to the meat, dairy, or egg industries.

U.S. District Court Judge James Robertson ruled on September 30, 2000, that the USDA violated federal law by keeping secret certain documents used in setting federal nutrition policies and by hiding financial conflicts of interest of the members of the Dietary Guidelines Advisory Committee. However, by the time of the ruling, the USDA had made the documents available to the public. The judge nevertheless reached an important legal decision with regard to balancing the public's right to information with the right of privacy of government advisors: "The asserted public interest is in learning whether a Committee member was financially beholden to a person or entity that had an interest in how the Dietary Guidelines might be amended. I find that public interest outweighs the privacy interest of the individual whose disclosure was redacted [deleted of personal information]."[22] This legal decision was narrowly focused and therefore does not create a general precedent for setting a standard of transparency for conflicts of interest among members of federal advisory committees.

New York University nutritionist Marion Nestle wrote in *Food Politics* that in her field of nutrition science and policy "co-opting experts—especially academic experts—is an explicit corporate strategy."[23] According to Nestle, companies screen the intellectual landscape for leading experts, court them, offer them research grants, and hire them as consultants. The experts slowly internalize the interests of their benefactors but retain their heartfelt belief in their own objectivity. Nestle muses, however, that because drug company sponsorship of research is highly correlated with publications of articles favorable to the sponsor's products, there is every reason to suspect similar biases are showing up in the field of nutrition. She believes that conflicts of interests are common among scientists serving on nutrition advisory committees, both in government and in the National Academy of Sciences. To the discomfiture of many of her colleagues, Nestle also documented the extensive network through which food companies sponsor nutrition research and the activities of professional societies. The major food companies sponsor nutrition conferences or make sizable contributions to scientific publications.[24]

EPA'S SCIENCE ADVISORY BOARD

The year 1989 will be recorded in regulatory history as the year of the great Alar controversy. Alar is the trade name of a growth-regulating compound whose chemical name is daminozide. It was first put on the market over twenty-five years ago and was used extensively on apple orchards to produce a homogenous color and promote uniform ripening.

While in the process of reevaluating the status of Alar as an approved pesticide, the EPA undertook an internal peer review by its own scientists. In August 1985, the EPA's peer review committee completed its analysis of the data on daminozide and its primary metabolite, UDMH. The committee concluded that the pesticide should be classified as a probable human carcinogen. The EPA then drafted a document, which it submitted to the agency's

Scientific Advisory Panel (SAP) for review, as required under the amendments to Federal Insecticide, Fungicide, and Rodenticide Act of 1947 (FIFRA). Later that year, the SAP criticized the results of the EPA's internal review that daminozide was a probable human carcinogen and claimed that this finding could not be supported by the existing data. Moreover, the panel concluded that the current science does not justify removing the pesticide from the market. In response, the EPA slowed down its initiative to restrict the use of Alar and to lower the allowable residue levels.

Meanwhile, the Natural Resources Defense Council (NRDC), one of the nation's leading environmental groups, released a report on February 27, 1989, titled "Intolerable Risk: Pesticides in our Children's Food." The report estimated the cancer risks to children from twenty-three pesticides. One of these pesticides was daminozide. The NRDC's risk assessment for this pesticide reported that its primary metabolite, UDMH, would cause one extra cancer for every 4,200 children exposed during their first six years of life.

The NRDC hired a public relations firm, Fenton Communications, to disseminate and promote the study. Fenton gave the CBS television news magazine *60 Minutes* an exclusive on the story. Among the twenty-three pesticides, the one CBS chose to focus on for its quarter-hour segment was Alar, considered among the most carcinogenic of the twenty-three pesticides. The *60 Minutes* segment, titled "A is for Apple," was seen by an estimated 40 million people. Within weeks, there was a nationwide boycott on apples and apple juice. When the consumer boycott grew large enough, food chains demanded non-Alar products. Supermarkets displayed signs with the "Alar-free" symbol (the word "Alar," with a line through it).

Apple growers and environmentalists were pitted against one another. The CBS television network was criticized by Uniroyal, the sole manufacturer of daminozide, as well as other industry-supported organizations for presenting an unbalanced characterization of the pesticide and for not reporting its benefits to farmers. After government hearings, a member of the House learned that seven out of the eight members of EPA's SAP for Alar had consulting relationships with Uniroyal, and at least one relationship was current. Public Citizen and the NRDC brought suit against the EPA for its practice of not disclosing the conflicts of interest of its scientific advisory panels. The suit did not result in full transparency of EPA science advisers.

Like other federal agencies, the EPA claims it seeks the most informed experts to serve on its science advisory panels. Quite often, these experts either work for or consult for industry. Federal agencies claim that if they excluded experts because they consult for industry or if they required disclosure of their private affairs, they could not fill the slots on their advisory panels.

GENERAL ACCOUNTING OFFICE INVESTIGATION OF EPA PANELS

Congress established the Science Advisory Board (SAB) of the EPA in 1978. The board comprised more than one hundred technical experts outside of government to provide science and engineering advice to the agency. The SAB convenes Scientific Advisory Panels (SAPs) of experts to review the rationale underlying the EPA's proposed regulations. One of these SAPs changed the course of the Alar regulatory debate, which eventually resulted in the high-visibility public controversy over the carcinogenicity of daminozide.

In June 2001, the General Accounting Office (GAO) issued a report on EPA's Science Advisory Board panels.[25] Peer review panels established by EPA's SAB must comply with federal financial conflict-of-interest statutes. Members of science panels are treated like "special government employees" (SGEs). Federal statutes prohibit federal employees from "acting personally and substantially in a particular matter that has a direct and predictable effect on their financial interests; those of their spouse or children or those of organizations with which they have certain associations, including employers."[26]

In some cases, the EPA chooses to have employees of industry on panels to provide a diversity of viewpoints and perspectives. The law does provide for exemptions and permits employees to participate in particular matters that have "direct and predictable effects" on their employer's financial interests. The exemption does not include the personal financial interests of the panelists in the particular matter, such as stock ownership in the employer.

As with the FDA and other agencies that require external scientific advice, the EPA has broad latitude to waive the conflict-of-interest provisions of the law. If the agency determines that the need for the special government employee's services outweigh the potential for a conflict of interest, it then has the authority to waive the conflict of interest.

The GAO study cited a number of problems with the way the conflict of interest provisions are implemented at the EPA. First, they reported that one-third of the financial disclosure forms filed by advisory panel members were not reviewed by the staff. This omission could thus allow individuals with conflicts of interest to serve as panelists without the agency's recognition of the scientist's personal financial relationships. In other words, no steps would be taken to either mitigate the conflicts or, where appropriate, grant the waiver.

Second, the GAO found that the disclosure forms are inadequate for eliciting a peer reviewer's conflict of interest: "A panelist on the revised cancer risk guidelines panel reported a prior long-term affiliation with a chemical industry organization that had commented to EPA on its revised guidelines."[27]

The forms may neglect to inquire about historical financial associations by focusing on current activities.

Third, the GAO concluded that the public is largely kept ignorant of what appears on conflict-of-interest forms submitted by EPA panel members. The EPA holds public disclosure sessions for discussing conflicts of interest among members of its Science Advisory Panels. The minutes of the disclosure sessions are made public. In its review of one panel's minutes (a review of the carcinogenicity of 1,3-Butadiene, hereafter known as the "1,3-Butadiene panel"), the GAO found that the disclosure forms left out important information germane to conflict of interest. For example, the minutes of the disclosure session revealed that two panelists owned stock in companies that manufacture 1,3-Butadiene and that two panelists received fees from chemical companies.

The EPA is responsible for ensuring that a diversity of perspectives are represented on the panels. The 1,3-Butadiene panel consisted of fifteen scientists: ten were professors, medical directors, or both at academic or medical institutions; four worked for companies; and one worked for a state environmental protection agency. According to the GAO study, the staff director of the SAB determined that six out of the fifteen chosen experts began their participation on the panel with an industry perspective on the issue of 1,3-Butadiene's carcinogenicity. Another panelist from an academic setting, who does exhaustive consulting for industry, was said to reflect the middle spectrum on the issue of concern. This finding is an indication that the agency's a priori determination of the panel's balance may understate the actual conflicts of interest of panel members.

The GAO's conclusions in its study are twofold. First, the EPA does not identify and mitigate conflicts of interest in a timely manner. Second, the public is not adequately informed about the special interests held by panel members when they enter the advisory process.

Government agencies, by and large, are not seeking to fill advisory positions with scientists that have conflicting interests. They claim that academia has been compromised. If government were to hold to a high ethical standard in selecting advisers, agencies contend that there would be a dearth of experts to fill the slots. That statement is essentially the dilemma. When selecting experts, choose either high ethical standards or high scientific standards—but you cannot have both. The leading experts are more likely to have commercial relationships. It is like the Heisenberg uncertainty principle, but applied to ethics and science.[28] Perhaps the standards for COI are unrealistic. What form do the conflicting interests among professors take, and how serious do they become? The next chapter examines the trend of dual-affiliated academic scientists who have learned how to use their position for developing a parallel career.

NOTES

1. Susan Wright, *Molecular Politics* (Chicago: University of Chicago Press, 1994) 529, n. 44.

2. U.S. General Services Administration, "Twenty-Seventh Annual Report on Federal Advisory Committees: Fiscal Year 1998" (March 1, 1999), 1, at policyworks.gov/mc/mc-pdf/fac98rpt.pdf (accessed February 20, 2001).

3. U.S. General Services Administration, "Twenty-Seventh Annual Report," 7.

4. U.S.C. App. 2 Sec. 5(b)(3).

5. See 18 U.S.C. 208(a), sec. 18 U.S.C., sec. 208(a), purpose.

6. President William J. Clinton, Executive Order No. 12838, "Termination and Limitation of Federal Advisory Committees," December 31, 1993.

7. Office of Government Ethics, "Interpretations, Exemptions and Waiver Guidance Concerning 18 U.S.C. (Act Affects a Personal Financial Interest)" *Federal Register* 61 (December 18, 1996): 66829.

8. 5 C.F.R 2640.103(a).

9. Anonymous, "How the Study Was Done," *USA Today,* September 25, 2000.

10. Dennis Cauchon, "Number of Experts Available Is Limited," *USA Today,* September 25, 2000, 10A.

11. Dennis Cauchon, "FDA Advisers Tied to Industry," *USA Today,* September 25, 2000.

12. Cauchon, "FDA Advisers Tied to Industry."

13. Cauchon, "FDA Advisers Tied to Industry."

14. Committee on Government Reform, U.S. House of Representatives, *Conflicts of Interest in Vaccine Policy Making,* majority staff report, August 21, 2000, at house.gov/reform/staff_report1.doc (accessed May 28, 2002).

15. Committee on Government Reform, *Conflicts of Interest*, 1.

16. Dan Burton, Chair, Committee on Government Reform, U.S. House of Representatives, letter to Secretary of Health and Human Services, Donna E. Shalala, August 10, 2000.

17. Committee on Government Reform, *Conflicts of Interest*, 19.

18. Committee on Government Reform, *Conflicts of Interest*, 27.

19. The FDA's Vaccine and Related Biological Products Advisory Committee (VRBPAC) met December 12, 1997, to vote on "Rotashield" as a vaccine for rotavirus.

20. Committee on Government Reform, *Conflicts of Interest*, 37.

21. Rose Gutfeld, "Panel Urges FDA to Act on Adviser Bias," *Wall Street Journal,* December 9, 1992, B6.

22. *Physicians Committee for Responsible Medicine v. Dan Glickman, Secretary, USDA.* Civil Action No. 99-3107. Decided September 30, 2000.

23. Marion Nestle, *Food Politics* (Berkeley: University of California Press, 2002), 111.

24. Marion Nestle, "Food Company Sponsorship of Nutrition Research and Professional Activities: A Conflict of Interest?" *Public Health Nutrition* 4 (2001): 1015–1022.

25. U.S. General Accounting Office, EPA's Science Advisory Board Panels (Washington, D.C.: GAO-01-536, June 2001).

26. U.S. General Accounting Office, EPA's Science Advisory Board Panels, 5.

27. U.S. General Accounting Office, EPA's Science Advisory Board Panels, 13.

28. The Heisenberg uncertainty principle in quantum mechanics states that it is impossible to measure the position and simultaneous velocity of a subatomic particle with unlimited precision. The measurement precisions of position and velocity are inversely related.

7

PROFESSORS INCORPORATED

When I first decided to pursue a career in academia, it was for three reasons. First, I had a passion for learning. I believed that if my life were centered in higher education and research, I would never get bored. A few summer jobs gave me a taste of routinized white-collar labor, and it didn't feel right regardless of the salary scale. The second reason was that I had a great appreciation for personal autonomy and self-direction. A career that allowed me to follow intellectual pursuits of my own choosing, rather than those dictated by others, would provide a unique sense of personal fulfillment and self-actualization. Finally, I liked the idea of a secular–monastic life that focused on ideas and inquiry and where the norms of the marketplace did not rule. Some of the most esoteric and inaccessible but significant works of scholarship, which have little if any market appeal, eventually achieve high esteem in the culture of the academy. Works of careful scholarship are not a slave to fashion. They rarely make it onto best-seller lists, but they do have a lasting place in our pantheon of collected knowledge. People slave over their works. The sense of personal pride in mastering a field or participating in a significant discovery is inestimable. Neither the scant royalties nor the academic salary can possibly provide a fair compensation for the countless hours of labor that go into producing a book of scholarly notoriety or in solving a mystery of nature.

For those who have the opportunity to receive a university education, they know that it is a right of passage to begin a career in an institution that looks and acts considerably different than does any alma mater. Our finest universities consist of highly decentralized academic departments that establish curriculum and set academic standards. Professional societies and the vast network of academic journals define acceptable scholarship and certifiable knowledge. Unlike many institutions, where critical choices are determined by market forces, decisions at universities are buffered against the pressures of supply and demand. There are no bottom lines or productivity measures that account for the qualities of a great university.

How would we view universities if they became the handmaidens to market-based institutions? Let us imagine that a career in higher education was used primarily to leverage personal wealth. In the language of the economist, there is unrealized value in the professoriate. Shouldn't the salaried professors use their social status to create additional wealth for themselves and their institutions? What changes in the public perception of universities would result when the pursuit of truth and the pursuit of wealth become equal factors in the incentive structure? Some of these changes in the professoriate are already underway, particularly in the fields of knowledge that can be easily turned into economic value. The consequences, thus far, have been mixed. This chapter focuses on some trends in the controversy over corporate professors.

The changes in institutional behavior that we now accept as the norm of university culture began to appear most visibly in the 1980s—a decade in which supply-side economics made its debut in Washington. The philosophy of government changed from its role in creating public wealth and providing a safety net for the less fortunate to a role in stimulating the expansion of the private, for-profit sectors. The new philosophy of government was to eliminate the impediments of its own doing to economic growth and competitiveness while maximizing personal choice and responsibility.

Concurrent with the changes taking place in economic policy—the transition from the Age of Roosevelt to the Age of Reagan—a revolution was in the making in the biological sciences. The discovery of gene splicing transformed biology into an untapped reservoir of economic potential for the creation of new drugs and medical assays, new clinical techniques in the treatments of disease, new methods of manufacturing, and new materials. Gene engineering changed biology from a predominantly analytical field, one that studied the structure and properties of living things, to a synthetic field, one that was capable of creating novel life forms that could be harnessed for human use. The subfield of biology that provided the tools for this transformation was molecular genetics. Its scientists discovered the chemical enzymes that cut and splice gene segments, and they developed the methods for deciphering the genetic code—the DNA composition—of any living entity. Like the physicists and engineers who broke the sound barrier and created artificial elements that were incorporated in the periodic table, biologists broke species barriers by transferring DNA segments across broad stretches of the phyla of living organisms.

Up until the time gene transplantation techniques were developed, a deep cultural divide existed between pure and applied biology. In the land-grant agricultural colleges, applied biologists used the knowledge of classical Mendelian genetics to select desirable traits in agricultural crops and food-source animals; pure biologists, however, investigated cell structure, including DNA composition, plant metabolism, and plant–environment interactions. Departments of applied biology had close ties with the agribusiness, particularly in areas of traditional biotechnology, which includes developing more efficient food and drink fermentation processes, as in the use of microorganisms like yeast to make beer. The division between pure and applied biology withered away when the discoveries in molecular genetics became recognized as broadly commercializable. Any scientist who could move genes around became instantly valuable to the emerging sector called "biotechnology." Gene splicing was the bridge that united the pure and applied divisions of biology. A scientist who spent his or her career investigating the genetics of the coliform bacterium *E. coli* became extremely valuable to a pharmaceutical company that saw in the microorganism the potential to produce limitless supplies of human proteins. Thus, pure biologists who recognized the commercial value of their work began starting their own companies or consulting for newly formed biotechnology firms.

In March 1981, *Time* magazine heralded the new industrial age of biotechnology with a cover story titled "Shaping Life in the Lab," which featured the disembodied head of millionaire scientist Herbert Boyer emerging from the strands of a cell's DNA. Boyer, a professor at University of California, San Francisco, started Genentech, one of the first generation of biotechnology companies established to develop new pharmaceutical products through genetic engineering. Hundreds of embryonic biotechnology firms, created from venture capital, formed symbiotic relationships with major universities. These firms sought the newly trained graduate students and postdoctoral researchers to work on company projects. They also needed eminent scientists to serve on their advisory boards, providing the intellectual "capital" that was essential for attracting the venture capital, high-risk investments.

In their masterful work *Academic Capitalism,* Slaughter and Leslie examine the new entrepreneurship of professors and their institutions, noting that several disciplines participated in turning the academy into private enterprise zones: "Biology was not the only basic science that became entrepreneurial and whose faculty lost their relative insulation from the market. In the 1990s a variety of interdisciplinary centers and departments developed—materials science, optical science, cognitive science—which became involved increasingly with market activity."[1]

But unlike the electrical engineering faculty in the 1950s and 1960s who left the university to start their own companies during the early days of the microelectronics revolution, scientific leaders in the fields of microbial, plant, animal, or human genetics (as well as other applied sciences) for the most part maintained their university appointments—as they became founding members, held equity, or served on scientific advisory boards of new companies. The impact of these dual-affiliated relationships on universities and the integrity of science were questioned in prestigious journals as early as 1980.[2] However, it took nearly a decade before the extent of these academic–commercial linkages was more fully understood.

In the mid-1980s, during a talk I delivered at MIT on the moral consequences of academic–industry ties, I posed a rhetorical question: What would a national linkage map look like if it were to show the affiliations biomedical scientists had developed with the rapidly growing biotechnology industry? If the number of participating scientists were small, then the impact on the university culture might be insignificant. If, however, it became the norm for academic scientists to be pursuing commercial interests while advancing knowledge, then the pillars of objectivity on which academic science has secured its moral standing might be vulnerable to collapse. Would there be a large enough reservoir of impartial scientists to protect the independence of academic science?

SCIENTISTS WITH DUAL AFFILIATIONS

Several years later, I decided that some of these questions could be answered. With the help from colleagues at Tufts University and a Harvard undergraduate, I developed a database of new biotechnology firms, which amounted to several hundred by the late 1980s. Most of the firms that issue publicly traded stock published the names of members of their scientific advisory boards either in company literature or in federally mandated reports. Private firms that do not file public reports were surveyed for their management personnel and scientific advisers. For the period 1985–1988, we found that faculty at leading research universities were already deeply invested in the commerce of biotechnology. Some of the data were reported in the journal *Science*.[3] At MIT, 31 percent of the faculty members in the biology department had formal ties to a biotechnology company. At Stanford and Harvard, about 20 percent of the biotechnology scientists had dual affiliations. The database of scientists we had compiled over a short period of time during the early stages of the commercialization of academic biology consisted of 832 biomedical and agricultural scientists who had formal ties to companies. At Harvard, the faculty ties involved 43 different biotechnology firms while at Stanford, MIT, and UCLA, it was 25, 27, and 19, respectively. This meant, in effect, that scientists at the same institutions were working for competing firms, and it therefore became commonly understood that the sharing of information was not to be taken for granted. One can easily understand the self-imposed constraints on communication arising out of faculty allegiance to forty-three companies, each seeking to protect its own intellectual property rights. Scientists at one leading research university reported that their colleagues "were loathe to ask questions and give suggestions in seminars or across the bench for there was a feeling that someone might make money from someone else."[4]

A survey of approximately 800 biotechnology faculty (published in 1992) revealed that 47 percent consulted with industry; nearly 25 percent received industry-supported grants or contracts; and 8 percent owned equity in a company whose products were related to their research.[5]

Many universities have turned a blind eye toward the entrepreneurial activities of faculty, preferring instead to pay homage to the valuable overhead funds that accompany these interlocking relationships. Two researchers studied faculty–industry relationships at their own university, the University of California at San Francisco (UCSF). Faculty members at UCSF are required by federal and state regulations to file conflict-of-interest disclosure forms. The investigators reviewed the forms filed between 1980 and 1999 and found a threefold increase of principal investigators who had personal financial ties to their industry sponsors (from 2.6 percent in 1985 to 7.1 percent in 1997). They also found that 32 percent of the principal investigators of sponsored research had held paid positions either on a scientific advisory board or on a board of directors of the company or agency from which they received their research funding. The authors stated: "Complex relationships, such as founding a company, serving on the advisory board, and owning stock, were not unusual and were viewed as problematic, though not completely unacceptable."[6] In 2001, the journal *Nature* reported in an editorial that "one third of all the world's biotechnology companies were founded by faculty members at the University of California."[7]

It has become a generally accepted norm at American universities that faculty are consulting in the private sector, serving on industry scientific advisory boards, receiving corporate honoraria, developing licensing agreements and patents, and participating in start-up companies. Harold Shapiro, who held presidency positions at Princeton and the University of Michigan, commented in a *New York Times* interview: "I don't think there has ever been a time in the history of modern scientific research when such a large proportion of those engaged in academic biological research are so involved with for-profit biotechnology companies. . . . It is hard to find scientists who are not potentially conflicted by their financial interests in these companies."[8] Can this pattern of behavior continue without any detriment to science? An editorial contributor to the *Journal of the American Medical Association* wrote: "When an investigator has a financial interest in or funding by a company with activities related to his or her research, the research is lower in quality, more likely to favor the sponsor's product, less likely to be published, and more likely to have delayed publication."[9]

In chapter 9, I address the question of whether the source of funding or the partnerships between academia and industry can introduce bias into scientific inquiry. Before we get to that question, however, I would like to pose another one: How can we assess the degree to which university faculty members have financial interests in their published studies?

AUTHOR CONFLICTS OF INTEREST IN PUBLISHED RESEARCH

Several years ago, L. S. Rothenberg (of the UCLA Medical School) and I decided to collaborate on a study to answer the following question: What is the likelihood that a lead author in a highly respected science or medical journal (usually the senior investigator) has a financial interest in the subject matter of the article? In other words, how many conflicts of interest are appearing in research publications, and when they exist, are they being disclosed?

One of the undisputed truths about research, whether in the social or natural sciences, is that it is far easier to pose a question than it is to answer it. Sometimes it is difficult, if not practically impossible, to acquire good data to resolve an issue. Other times, it is simply too expensive to carry out a study. To answer the question we posed, we first needed to develop criteria for the phrase "possessing a financial interest." We did so because one of the first principles of research is that you cannot measure what you do not define. Second, we had to develop techniques for determining whether the journals' authors met our definition of "possessing a financial interest" in their publications.

Fortunately, professional and federal guidelines exist that provide examples of the circumstances under which scientists should disclose their financial interests related to their research. The problem we faced was that we could not get measurements for all of the interests a scientist might have. Some, like stock holdings and honoraria, were out of bounds for any study except one that would survey the individual scientists. To find the answer to our query, we thus decided to use available objective information, as opposed to using traditional subjective responses to a survey questionnaire.

We selected three disclosable interests of academic scientists that we could measure: serving on a scientific advisory board of a company that develops products in an area related to the scientist's research; possessing a patent or a patent application for a discovery or invention closely related to the scientist's work; and holding a position as an officer or a major stockholder in a company whose products are related to the scientist's research. The details of the study are given in chapter 10.

While we expected to find scientists with financial interests in their research, we were not prepared for the high level of involvement that we discovered. Nearly 34 percent of the articles (267) had at least one lead author with one of the financial interests we measured. In other words, if you picked out an article at random authored by a Massachusetts-based scientist from one of these fourteen journals, the chances are one out of three that one of the lead scientists had a financial interest in its outcome. We also learned that none of the lead scientists disclosed their financial interests. Such a discovery is not surprising because in 1992 most of the journals did not have mandatory conflict-of-interest disclosure policies for authors.[10] Chapter 10 addresses the nature and effectiveness of journal policies on conflicts of interest.

While we cannot say with any degree of certitude that the level of commercial involvement of Massachusetts academic scientists was similar to that of scientists in other states, our findings (even if they represented double the national average) would be troubling. Regardless of the existing level of deviance in science (misconduct, bias, violation of ethical norms), the introduction of commercial values into the research enterprise will certainly exacerbate it. The fact that one-third of the articles we sampled had authors with a financial interest in the subject matter tell us that business and academic science have merged. The consequences of this merger are not fully understood. What we do know, however, is that new variables must be added to the study of the norms and motivations of academic scientists.

APPLYING THE UNIVERSITY LABEL

How far away are we from certifying products with a university label? Presumably, only the trustees of a university can allow their logo or good name to be used on behalf of some private venture. By and large, the idea behind selling the university's imprimatur on behalf of some commercial item has not been overtly accepted. Yet we do have examples where faculty have used their university titles and affiliations to support a product that, in the public's mind, is easily translated as an institutional endorsement.

As reported in the *News & Observer* (Raleigh, North Carolina), an associate professor at North Carolina State University (NCSU) developed a method to dispose of chicken carcasses. This is a major problem for the poultry industry, which has to destroy thousands of dead or sick carcasses annually. This professor set up a company to market his idea. A company brochure stated that the technique was developed by an NCSU professor and that the process satisfies "the performance specifications determined by the NCSU Poultry Science Department."[11] On investigation, the reporters for the *News & Observer* learned that no department performance specifications even existed. On discovering that the university's name was improperly used for advertising purposes, the remaining brochures were destroyed.

Unfortunately for NCSU, this case was not an isolated one. Another NCSU professor, a weed scientist who did consulting work for the pesticide producer Rhone-Poulenc, appeared in a brochure produced by that company (and another one, Stoneville Seed), endorsing a new genetically modified plant that is herbicide tolerant.[12]

According to the Association of University Technology Managers, more than 180 universities own stakes in 886 start-up companies. University names are increasingly associated with certain products. Consider the ones that are now part of the American lexicon of commercial products. The University of Florida created Gatorade; UCLA developed the nicotine patch; and Stanford University students created the Yahoo! and Google search engines. Even when the university is not mentioned as an endorser of a product, faculty members with a university title are themselves endorsing products. For example, Professor Stanley Malamed, chair of Anesthesia and Medicine at the University of Southern California School of Dentistry, developed a dental device called "The Wand," which is designed to deliver a "painless dose of anesthesia." Professor Malamed, while a consultant to a biomedical company called Milestone Scientific, endorsed "The Wand" without ever discussing its limitations.[13]

Universities have been cautious about allowing their names to be used to promote products or medical techniques. The autonomy and freedom of association afforded to professors, however, has meant that they are for the most part free to endorse commercial enterprises using their institutional affiliation, thereby establishing some tacit connection between the university and the enterprise. When the name of a professor from a prestigious university appears on a company brochure, it projects an image of integrity that transfers from the institution to the company and its products. The danger for the institution is that one scandalous product could permanently tarnish its reputation or drain its resources in a product-liability suit. Slaughter and Leslie warn of the economic consequences to a university that has to defend itself against a blockbuster legal challenge to products like thalidomide or breast implants.[14]

GHOST AUTHORS

Few taboos in academic life elicit the universal condemnation more than plagiarism does. Many universities have guidelines on academic integrity that are distributed to incoming students. Under the section regarding plagiarism, my own university's guidelines state: "The academic counterpart of the bank embezzler and of the manufacturer who mislabels his products is the plagiarist, the student or scholar who leads his reader to believe that what he is reading is the original work of the writer when it is not."[15]

In professional life, things get more complex. For example, famous personalities hire ghostwriters to write a book. Sometimes their names do not even appear on the book's cover. They get no attribution, only a paycheck. Likewise, it is commonly known that U.S. presidents and other busy, high-profile public officials have speechwriters. Someone wrote John Kennedy's words "Ask not what your country can do for you, ask what you can do for your country" and the words he spoke in West Germany that brought tears to Berliners: "*Ich bin ein Berliner.*" Yet, by convention, we attribute the ghostwritten words to the person who spoke them and not to the person who first penned them.

A presumption exists regarding copyright law that the person who pens the words has a copyright for what was written. Of course, that copyright must be asserted and protected from stealth and plagiarism. However, within copyright law, provisions are created for "work for hire." If someone contracts a story or study from a writer, then the writer may not have the copyright and thus may not be able to declare authorship. The contractor in "work for hire" owns the copyright and may rightfully declare ownership. In "work for hire," the story may be altered, and it generally would not appear with the original author's name. Similarly, most daily newspapers retain the copyright of articles produced by their staff reporters, even when the staff reporters are given a byline. But newspapers do respect authorship to the extent that they would not replace the authentic author's name with that of someone who did not write the story.

When we see a scientist or medical researcher listed as the author of an article, can we assume, therefore, that it was written by that individual? It may surprise some people to learn that there is a ghostwriting industry in science and medicine. This industry is not targeted to students, but rather to professional scientists and medical researchers. In April 2000, an investigative story published in the *Hartford Courant* disclosed the following: "In 1994, Wyeth [a pharmaceutical company] signed a $180,000 contract with a New Jersey medical publishing company called Excerpta Medica that offered pharmaceutical companies an invaluable tool: ready made scientific articles placed in leading medical journals, and carrying the imprimatur of influential academic leaders."[16]

It works like this: Excerpta is contracted by a company to find a distinguished academic scholar to agree to have his or her name placed on a commentary, editorial, review, or research article, which has been written by someone either from the company or someone selected by Excerpta. The practice is defended by the rather spurious argument that the academic whose name gets placed on the article always has final editorial review. Such practices would violate the minimal standards of plagiarism we demand of our students. In the case reported by the *Hartford Courant*, the writer of the article was a freelancer who was paid $5,000 to research and write the article under company standards. The university scientist, whose name appeared as the author, was paid $1,500.

Representatives of the pharmaceutical industry claim that it is a common practice to have articles that appear in journals ghostwritten by freelancers. Some medical and scientific journals ask authors to certify that each person listed on the article had a substantial role in the research study or the written product. The *Journal of the American Medical Association* requires its authors to sign the following statement to meet its criteria for authorship. "I have participated sufficiently in the conception and design of this work and the analysis of the data (when applicable), as well as the writing of the manuscript, to take public responsibility for it."[17]

In their book *Trust Us, We're Experts*, Sheldon Rampton and John Stauber wrote: "Pharmaceutical companies hire PR firms to promote drugs. . . . Those promotions include hiring freelance writers to write articles for peer reviewed journals under the byline of doctors whom they also hire."[18] If ghostwriting is so prevalent in science and medicine, then why haven't professional societies and journals discredited it as a form of scientific misconduct?

Sarah Bosely, the health editor of the *Guardian Weekly* (UK), reported on the increasing prevalence of ghostwriting in medicine. Citing survey results, Bosely wrote that the going rate for British psychiatrists to undersign a ghostwritten article was $2,000, and for American psychiatrists, the rate starts at $3,000 and goes as high as $10,000. "Originally, ghostwriting was confined to medical journal supplements sponsored by the industry, but it can now be found in all major journals in relevant fields. In some cases, it is alleged, the scientists named as authors will not have seen the raw data they are writing about—just tables compiled by company employees."[19]

In May 2002, the *New York Times* broke a story about physicians who received money in exchange for allowing pharmaceutical sales representatives into their examining rooms. Warner-Lambert (subsequently acquired by the drug giant Pfizer) was experimenting with new marketing strategies for expanding off-label uses of a drug approved for epilepsy called Neurontin. Unsealed court documents revealed that "Warner-Lambert also hired two marketing firms to write articles about the unapproved uses of Neurontin and found doctors willing to sign their names to them as authors."[20] Warner-Lambert paid the firm $12,000 per article and $1,000 to each doctor willing to serve as an author.

It is not clear whether ghostwriting or honorary authorships are generally considered violations of scientific norms. Some authors are not embarrassed to admit their involvement in these activities. Journals differ on their level of tolerance or abhorance of these perversions of authorship.

So how much of this practice is actually taking place? A team of investigators tried to answer this question for a group of medical journals.[21] They surveyed 809 authors of general medical journals, all of which followed the guidelines for authorship issued by the International Committee of Medical Journal Editors (ICMJE), a voluntary group of editors who have given special attention to the ethics of publication. They used a computer-generated random number list to sample the articles published in the journals during 1996. Of the 809 articles sampled in the six medical journals, they found that ninety-three were ghost authored (11 percent). This broke down to 13 percent research articles, 10 percent review articles, and 6 percent editorials. With such sizable percentages, it is difficult to view ghost authorship (or honorary authorship for that matter, which had a 19 percent prevalence in the study) as aberrations within the norms of scientific publication.

If a ghostwritten article is an example of phantom authorship, the next level of deceit involves the fabricated credentials of scientists. Like ghost authorship, many of these cases only see the light of day when there is litigation and when the case goes to trial or the court documents are unsealed.

VIRTUAL SCIENTISTS AND THE COURTS

In 1999, I was contacted by a Chicago law firm and asked if I would consider being an expert witness for a corporate plaintiff. They were interested in me because the case involved ethics and scientific integrity. I was sent a formidable stack of documents and the complaint that was filed with the United States District Court for the Northern District of Georgia, Atlanta Division. The documents that I reviewed were sufficiently provocative and (assuming they were credible) illustrative of questionable practices between the corporate defendant and various academic researchers. For that reason, I accepted the invitation. My role was to read the documents and situate the practices they described within the current ethical norms of science.

The case was brought to trial in early 2002, and some of the documents previously sealed under court order became publicly available. (The entire trial record is now a public record.) The case illustrates how a corporation engages in the following practices: create a funding strategy directed at academia to support the scientific claims behind one of its products; develop relationships with scientists; become involved in establishing research protocols; edit drafts of articles for university scientists; and help them find journal venues for favorable publications.

The plaintiff in this case was the Allegiance Healthcare Corporation, which manufactures and distributes a myriad of health care products, including natural rubber latex gloves. Allegiance filed suit against Regent Hospital Products, a subsidiary of London International Group and also a competitor in the health care products industry (another manufacturer of natural rubber latex gloves). Allegiance's complaint alleged that Regent made false and misleading advertising claims, which damaged its market share of latex gloves. The complaint stated: "Regent's promotional campaign for its natural latex gloves has largely become a campaign against the natural rubber latex gloves of other manufacturers."[22]

Allegiance's complaint included allegations of the gross misrepresentation of the educational credentials of a key employee of Regent, a woman by the name of Margaret Fay, who held the title in the company of Global Medical Affairs Director. The plaintiff argued in its opening remarks at the trial:

> Regent represented at one time or another to the public that Ms. Fay had a bachelor's degree from Columbia University in New York; that she had a Ph.D. from the University of Minnesota, Minneapolis, St. Paul; that she had done post doctoral studies in immunology at Cornell University and at other universities; that she was a professor of surgery at the University of Virginia Medical School. Regent claims that Ms. Fay was a member of the National Science Foundation, and the National Academy of Science, and the American Academy of Science. All of these claims are false. . . . None of them are true.

I spent considerable time examining the primary evidence for the claim that Margaret Fay's scientific credentials were grossly fabricated. The fabrication was especially relevant since Fay, as Global Medical Affairs director, was in charge of a $20 million external research budget. I became convinced that the evidence of the false representations of her scientific standing—which appeared in biographies, brochures, and résumés describing Fay, at times published in health or medical journals—was indisputable.

Investigations of her credentials revealed that there were no records of Fay ever having attended or having received a degree from Columbia University. Fay had some contact with Columbia Pacific University, a diploma mill that has been denied accreditation by the California Council for Private Post-Secondary and Vocational Education. This virtual institution specializes in distance learning and offers credit for life experiences. The evidence presented by the plaintiff showed that there is no record of Fay's ever attending or receiving a degree from the University of Minnesota, nor was there a record of her on the current or any previous faculty at the University of Virginia School of Medicine, including its Department of Plastic Surgery. She once had a title of "visiting research professor" at the University of Virginia Medical School, an honorary, unpaid affiliation that allows a person to do research in the university library. There was no record of her ever attending Cornell University or the University of Minnesota. Moreover, as anyone in the scientific community knows, the National Science Foundation does not have members. In addition, Fay was never elected to the National Academy of Sciences, one of the nation's most distinguished scientific societies. This complete fabrication of her scientific credentials (she did have a nursing degree) was admitted under cross-examination by Fay herself during the trial.

The case went to trial in February 2002. During the plaintiff's questioning of the executive director of Regent's North American business operations, the following exchange was recorded.

> *Q.* One of the things Regent did was play up her [Margaret Fay's] credentials, didn't they?
>
> *A.* Yes, Sir.
>
> *Q.* Because it's awful impressive to be a professor of plastic surgery at some place like the University of Virginia, isn't it?
>
> *A.* It was listed. [The title was listed on her credentials with the company's association.]
>
> *Q.* Now, as corporate representative of Regent, having spent, you know, I'm assuming a lot of time preparing for the trial, have you been able to find a single thing that Regent did to ever verify the credentials of Ms. Fay, who they held out to the world as being one of the top experts in this area.
>
> *A.* No, Sir.

Even after Regent found out about Ms. Fay's grossly fabricated credentials, it made the decision to hide that fact from everyone, including the researchers and universities Fay had worked with.

> *Q.* The decision was made at the highest executive levels at Regent and London International not to tell the world about Ms. Fay's true credentials, Is that right?
>
> *A.* Yes, Sir.

Could Ms. Fay and/or Regent have been complicit in this extraordinary fabrication of her credentials? Did this fabrication affect the collaborations of Ms. Fay with scientists at universities who received support from Regent? What type of funding, ghostwriting, research manipulation, and control might have resulted from the collaboration of a "constructed" scientist at a company who collaborated with other medical researchers, who might have been "taken in," or who were eager to receive their bonus checks?

During the trial, the head of Regent responded to questions about Fay's role in ghostwriting articles.

> *Q.* Sir, you are aware that Ms. Fay during her employment with Regent was ghost-writing articles to be published under researchers, other researchers' names?
>
> *A.* I was aware that she was ghost-writing articles.
>
> *Q.* You were aware, sir, as the highest officer of the defendants in the United States, that not only was Ms. Fay ghost-writing articles, Manning, Salvage and Lee, your public relations firm, was ghost writing articles for researchers?
>
> *A.* I believe that is correct, yes.

In her position with Regent, Fay financed dozens of academic scientists in the United States and Europe to test Regent's latex gloves for various attributes against its competitors. She helped to design testing protocols; she drafted and contributed ghostwritten articles; and she heavily edited manuscripts of scientific findings about her company's or its competitors' products written by authors funded by Regent. Fay's role was to provide the proper spin on research findings in published papers to ensure that her company's product looked as good as it could or that its competitors' products looked as bad as they could. Her judgments for changing the wording of a scientific paper were taken by the scientific authors with the gravitas that we are likely to extend to colleagues who have been through rigorous scientific training and who hold weighty degrees. The trust we have in the judgment of fellow scientists is based on the authenticity of their training and research background. In this situation, we have a high-ranking member of a corporation, with lofty scientific credentials all fabricated, who had money to fund research and other activities of academic scientists. Many studies by Regent were funded only under the condition that the company could stop publication if the results were in conflict with their marketing objectives. There was no evidence that any of these dubious activities was against the law, yet this behavior and corporate decision making precisely constituted the recipe for corruption and bias.

The jury's ruling on the case was largely in favor of the defendants, but it did decide for the plaintiff on two questions—namely, that Fay's credentials were false and that her credentials were used in commercial speech. However, the jury found that Fay's deceit caused no harm to Allegiance.[23] The jurors were apparently forgiving of Fay's deceit, which became lost in a struggle between two corporate entities vying for market share.

At this point, I have to ask: What message does this ruling send to corporations who feel justified in constructing their own experts and their own science?

• • • • •

My testimony on the ethics of science was never heard by the jury. The trial judge ruled in favor of Regent in connection with a motion it filed to prevent me from testifying on issues in the case that bore on scientific integrity. The judge accepted the argument based on a narrow interpretation of federal legal precedent (e.g., *Daubert v. Merrell Dow*)[24] that gives trial judges the discretion to restrict experts to those areas of dispute that require "true expert knowledge" (i.e., expertise grounded in the methods and procedures of science). In the case of my area of expertise, scientific integrity and ethics, the judge ruled that Allegiance's lawyers could speak directly to the jurors on this issue without the benefit of someone with scholarly credentials in the field. In addition, because violating norms of scientific integrity is not against any law, my testimony was viewed as irrelevant to whether Regent's conduct was unlawful.

From the media's perspective and the public's interest, this case was a low-profile one between two corporations battling for market share of their products (latex gloves) and contesting whether federal laws about marketing and advertising practices were violated or not. The behavior of a corporation in shaping research among academic scientists who support its products is part of the backdrop. But the easily forgotten details of this case provide a key to our understanding how easily science can be corrupted. When there are neither legal nor effective moral safeguards to protect the integrity of research, the manipulative corporate funding of science will surely exploit our university professors' conflicts of interest.

NOTES

1. Sheila Slaughter and Larry L. Leslie, *Academic Capitalism* (Baltimore: The Johns Hopkins University Press, 1997), 7.

2. The journal *Nature* published a debate between David Baltimore and myself on the academic's role in the commercialization of biology: "The Ties That Bind or Benefit," *Nature* 283 (January 10, 1980): 130–131.

3. Eliot Marshall, "When Commerce and Academe Collide," *Science* 248 (April 13, 1990): 152.

4. Quoted in David Dickson, *The New Politics of Science* (New York: Pantheon, 1984), 78.

5. David Blumenthal, "Academic–Industry Relationships in the Life Sciences," *Journal of the American Medical Association* 268 (December 16, 1992): 3344–3349.

6. Elizabeth A. Boyd and Lisa A. Bero, "Assessment of Faculty Financial Relationships with Industry: A Case Study," *Journal of the American Medical Association* 284 (November 1, 2000): 2209–2214.

7. Editorial, "Is the University–Industrial Complex Out of Control?" *Nature* 409 (January 11, 2001): 119.

8. Howard Markel, "Weighing Medical Ethics for Many Years to Come," *New York Times* July 2, 2002, D6.

9. Catherine D. DeAngelis, "Conflict of Interest and the Public Trust," editorial, *Journal of the American Medical Association* 284 (November 1, 2000): 2193–2202.

10. Sheldon Krimsky, L. S. Rothenberg, P. Stott, and G. Kyle, "Financial Interests of Authors in Scientific Journals: A Pilot Study of 14 Publications," *Science and Engineering Ethics* 2 (1996): 395–410.

11. Trish Wilson and Steve Riley, "High Stakes on Campus," *News & Observer,* April 3, 1994.

12. Steve Riley, "Professor Faulted for Role in Brochure," *News & Observer,* September 18, 1994.

13. Mathew Kaufman and Andrew Julian, "Medical Research: Can We Trust It?" *Hartford Courant,* April 9, 2000, A1, 10.

14. Slaughter and Leslie, *Academic Capitalism*, 203.

15. Tufts University, Dean of Students Office, School of Arts, Sciences, and Engineering, *Academic Integrity @Tufts,* August 1999.

16. Mathew Kaufman and Andrew Julian, "Scientists Helped Industry to Push Diet Drug," *Hartford Courant,* April 10, 2000, A1, A8.

17. See *Journal of the American Medical Association,* "Instructions for Authors" at ama-assn.org/public/journals/jama/instruct.htm.

18. Sheldon Rampton and John Stauber, *Trust Us, We're Experts* (New York: Tarcher/Putnam 2001), 201.

19. Sarah Bosely, "Scandal of Scientists Who Take Money for Papers Ghostwritten by Drug Companies," *The Guardian Weekly,* February 7, 2002.

20. Melody Peterson, "Suit Says Company Promoted Drug in Exam Rooms," *New York Times,* Business section, May 15, 2002, 5.

21. Annette Flanagin, Lisa A. Carey, Phil B. Fontanarosa, et al., "Prevalence of Articles with Honorary Authors and Ghost Authors in Peer-Reviewed Medical Journals," *Journal of the American Medical Association* 280 (July 15, 1998): 222–224.

22. *Allegiance Healthcare Corporation v. London International Group, PLC., Regent Hospital Products, Ltd., LRC North America.* United States District Court for the Northern District of Georgia, Atlanta Division. Civil Action No. 1:98-CV-1796-CC, pp. 2–3.

23. *Allegiance v. Regent.* United States District Court for the Northern District of Georgia, Atlanta Division. Civil Action No. 1:98-CV-1796-CC.

24. *Daubert v. Merrell Dow Pharmaceuticals, Inc.*, 43F.3d 1311 (9th Cir 1995).

8

CONFLICTS OF INTEREST IN SCIENCE

On almost any day of the year, stories appear in the print media that draw attention to conflicts of interest in association with some public official. For curiosity, I entered the keyword "conflict of interest" in the popular news and legal research database Lexus-Nexus on a rather ordinary day in August. My search revealed four stories for that day in major U.S. newspapers reporting conflict-of-interest allegations.[1] For the previous thirty-day period, Lexus-Nexus showed that nearly five hundred stories were published in national newspapers under the key word "conflict of interest."

The term "conflict of interest" is used like a flashing yellow signal to alert society to proceed with caution in the face of some actual or potential wrongdoing, primarily among people who hold positions of public trust. It is also meant to describe actions that should be taken to avoid a moral indiscretion, the appearance of wrongdoing, or a violation of law.

For instance, a vice presidential candidate establishes a blind trust for his investment portfolio so that, following the law, he would not make any policy choices that are consciously financially self-serving. A judge recuses herself from sitting in judgment of a case because a member of her family has a personal friendship with one of the litigants. A congressman who sits on an FDA oversight subcommittee refuses any campaign contributions from pharmaceutical companies and culls his investment portfolio of any individual pharmaceutical stocks. A national news anchorwoman discloses that the investigative story about to be aired contains allegations of fraud against the parent company of the network.

Conflicts of interest are ubiquitous in the organization of any social structure. The more complex the structure (i.e., the higher the number of interlocking relationships), the greater the potential for conflicts of interest. To claim that someone has a conflict of interest suggests that this individual may not be in a position to discharge his or her public responsibilities without partisanship because of conflicting relationships.[2] It is not meant, however, to apply to one's ideas or principles. Thus, we would not ordinarily invoke the term "conflict of interest" in connection with a death penalty pardon by a governor who happens to oppose capital punishment—although we certainly might hear the term applied if the governor's brother were paid handsomely by the family of the incarcerated felon to lobby for his pardon.

Our language has evolved a nomenclature specific to "conflict of interest," and it includes such terms as "self-serving," "nepotism," "abuse of power," "self-dealing," "insider trading," and "quid pro quo." In the United States, a sizable body of federal statutes and regulations are directed at preventing and penalizing public employees from using their office or position to derive personal wealth or benefit friends and family. These laws also apply to behavior after a person leaves elected office or a government position.

DECONSTRUCTING "CONFLICT OF INTEREST"

In his book *Conflict of Interest in Public Life*,[3] Andrew Stark provides an anatomy of conflict-of-interest behavior that reduces it down into three stages. The *antecedent acts* (stage 1) are factors that condition the state of mind of an individual toward partiality, thereby compromising the potential of that person from exercising his or her responsibility to foster public interests, rather than private or personal ones. The *states of mind* (stage 2) represent the affected sentiments, proclivities, or affinities conditioned by the antecedent acts. Thus, a politician who accepts a substantial campaign contribution from an individual may be more inclined to favor that individual's special interests than if no contribution were given.

The final stage represents the *outcome behavior* (stage 3) of the public official, or those actions taken (decision behavior) that arise from an affected state of mind as influenced by the antecedent conditions. The outcome behavior could result in self-aggrandizement or in rewarding friends at the expense of the general public interest.

The public official acts unethically (and sometimes illegally) when his or her behavior (stage 3) fails to measure up to a public interest standard and instead rewards those related to the antecedent acts (stage 1). In sum, the sequence of stages develops from the antecedent acts (such as financial involvements) to the states of mind, and finally, to the behavior of partiality.

[Antecedent Acts] → [States of Mind] → [Behavior of Partiality]

If public conflict-of-interest law were directed only at stage 3, the "behavior of partiality," this focus would have several implications. First, a person could be found guilty of a conflict of interest only if it could be proved that his or her behavior *resulted from* gifts, favors, or unsavory relationships. We cannot infer from decisions that we *believe* to be self-serving that they are in fact the consequence of questionable relationships a policy maker had with different stakeholders. Second, it is difficult in law to characterize a person's state of mind. We may know that a decision maker accepted a gift and that his or her decision favored the gift giver. But we do not know (and would have difficulty demonstrating) that the decision grew out of a state of mind of partiality toward the gift giver.

Consider a case where a U.S. president issues a pardon to a person, currently living outside the United States, who is charged with a felony and has never stood trial. Funds contributed to the president's campaign can be traced to the alleged felon's immediate family. How can it be shown that there is a link between the gifts, the president's state of mind, and the decision to issue a pardon?

The third implication of focusing exclusively on outcome behavior in conflict-of-interest law is that it would have little prophylactic effect. Most of the damage is already done by the time the legal processes kick in. Only a small number of cases would be prosecuted since the burden to demonstrate violations would be high.

On the grounds that "states of mind" are not amenable to legal remedies for conflicts of interest, the first stage (that behavior fosters a state of mind toward self-aggrandizement and partiality) has become the target of regulatory law. Stark notes that

> because we cannot prevent officials from mentally taking notice of their own interests, we prohibit the act of holding certain kinds of interests in the first place. Because we cannot prohibit officials from becoming mentally beholden to those who give them gifts, we prohibit the very act of receiving gifts under certain kinds of circumstances. Because we cannot prevent officials from being mentally influenced or ingratiating, we forbid them the acts of contacting official colleagues (former or current) or private employers (prospective and previous) in particular situations.[4]

He argues that the conflict-of-interest laws are preventive in nature by prohibiting actions of public employees that are antecedent to the development of states of mind that may be favorable toward special interests. A public official taken to lunch is hardly enough to influence public policy; nevertheless, the law proscribes these and other types of activities for public officials that could, in their aggregate, result in quid pro quo behavior.

America's brief but intense period of moral self-examination came after the Watergate affair, when an administration replete with corrupt appointees and an elected president who knowingly flaunted the law were purged. Several years later, Congress passed the Ethics and Government Act (EGA) of 1978. It was not as if the country had just discovered public ethics; rather, there was simply insufficient monitoring and accountability for the ethical transgressions in the form of public officials trading favors and making self-serving decisions.

The EGA (Title 1) requires public officials and candidates for office to disclose their financial assets and certain gifts that they have received. It establishes requirements for public officials to set up blind trusts for their assets when their position could result in selective self-aggrandizement. The EGA has a section (Title V) devoted to postemployment conflict of interest. It bans senior ex-officials from lobbying their own former agency for a year after leaving office, and it makes it illegal for ex-government employees to serve as a legal representative of a party that is litigating the government. In addition, it sets limits on the activities of law partners of members of the executive branch, and it mandates that the attorney general promulgate rules and regulations prohibiting officials in the Department of Justice from supervising investigations if they are in a conflict of interest.

Recusal is one of the methods of managing conflicts of interest among government officials and judges. It is used in cases where decision makers cannot reasonably shield themselves from the influencing factors. Instead, they shield themselves from making a decision. Recusal is a personal decision on the part of an official to avoid the aftermath of criticism and suspicion that often accompanies the disclosure of conflicting or self-aggrandizing interests in public decision making. Secretary of Defense Donald Rumsfeld, who under the administration of President George W. Bush, reported that he avoided making decisions involving weapons programs and AIDS policy "because of his potential conflicts caused by his once-vast stock portfolio."[5] Under the current conflict-of-interest law, Rumsfeld was required to replace his deferred compensation from the Kellogg Company, where he served as director, because the Pentagon purchases cereal from the company.

SCIENTISTS FACING CONFLICTS OF INTEREST

Public interest law that addresses conflicts of interest has clearly evolved over a period of years. But the association of conflict of interest with scientists and medical researchers is relatively new, and the management of those conflicts is still evolving. What are the conditions under which scientists could become encumbered with conflicts of interest? They can occur, according to Thompson, when a set of conditions exist "in which professional judgment regarding a primary interest (such as a patient's welfare or the validity of research) tends to be unduly influenced by a secondary interest, such as financial gain."[6] In a report issued by the Association of American Medical Colleges, the term "conflict of interest in science" was described as "situations in which financial or other personal considerations may compromise, or have the appearance of compromising an investigator's professional judgment in considering or reporting research."[7] An example offered by the *New England Journal*'s editor emerita, Marcia Angel, captures this point: "If an investigator is comparing Drug A with Drug B and also owns a large amount of stock in the company that makes Drug A, he will prefer to find that Drug A is better than Drug B. That is the conflict of interest." Angel notes that that the COI is a function of the situation and not the investigator's response to it.[8]

Other conflicting interests in science exist besides financial ones, and they have been addressed through a variety of mechanisms. For example, it is generally not acceptable for an academic scientist to serve as a reviewer on a grant proposal when the author of the proposal is from the same institution as the reviewer. The doctrine of rewarding friends and punishing enemies carries across the boundaries of science. Journal editors understand that there are deep intellectual divisions in science, which, at times, may account for the polarization of views about the quality of a submitted paper.

But intellectual divisions usually become apparent among scientific research groups. Considerations are given to the fairness of reviews when the potential exists that strongly held beliefs against certain theories can distort the process. These considerations, however, do not necessarily protect those who are solo believers and cannot find support from other researchers in their community of scientists. The two-time Nobel laureate Linus Pauling had difficulty getting funding for his theory about megadoses of vitamin C as a preventative of certain diseases.

Because scientists are rather new to the idea of "conflicts of interest," they have come to view them quite differently than those who participate in public life, where it has an established role in law and professional ethics. The typical scientist finds it incredulous that any financial interest they might have connected to their research would affect the way they do science. Scientists foremost view themselves as contributors to the frontiers of knowledge, whether pure or applied. Their primary commitment is to the discoveries they make and the applications of those discoveries to human society that may result.

As a consequence, most scientists view conflicts of interest as a public perception problem. They believe people draw inferences that scientists who have commercial interests in an area of research or are funded by a for-profit institution may carry a bias toward their research. For this reason, to protect the public's trust in science, many researchers have come around to believe that some sort of public-relations response to the rise of entrepreneurial science is warranted. Except in unique circumstances, where a scientist has abandoned his or her professional calling, it is generally considered anathema among scientists to purposefully alter their research goals or outcomes in the interests of a sponsor or their investment portfolio. In other words, it is widely accepted among members of the scientific community that the "state of mind" of the scientist is not prone to the same influences that are known to corrupt the behavior of public officials.

The prophylactic measures that are taken to prevent conflict of interest in public affairs are considered irrelevant in science precisely because scientists view themselves as participating in a higher calling than that of public officials—namely, the pursuit of objective knowledge. While senior public officials (elected or appointed) are prohibited from managing their portfolios during their tenure in office, scientists with patents and equity in companies that fund their research are at most simply asked to disclose their interests.

Some universities have established restrictions against certain potentially conflicting relationships. At Harvard University, for example, scientists who own more than $20,000 in publicly traded stock cannot serve as the principal investigator on a research grant funded by the same company. They are also restricted from receiving more than $10,000 annually in consulting fees or honoraria from companies that sponsor their research. If policies like this were to become the norm, they would help to dispel the myth that scientists are less vulnerable than are members of other groups to lapses in integrity.

It is a penalizable offense when public officials fail to disclose their conflicts of interest, yet for most scientists, where there are disclosure requirements, it is essentially voluntary. As we shall see in chapter 10, many of the journals that have adopted COI policies allow their authors not to disclose their financial interests, and some journals even fail to have penalties for authors who flaunt the policy.

Three principal areas exist where conflicts of interest intersect the world of science. First, clinical researchers may appropriate body tissue or genetic information as intellectual property from people who come under their care. The interest of a clinical investigator who is commercializing people's unique cell lines or genetic markers may come into conflict with their role as caregiver. Second, in clinical research, there is a concern that the physician's financial interests in an experimental treatment may compromise patient care by downplaying risks. Finally, the third concern is that the financial interests of the researcher or the sponsor may influence the outcome of the research results by influencing the way an investigator chooses to carry out the study or interprets its findings. The influence of the private sponsorship of academic research on the outcome is not easy to detect. The sponsor's interests may bias the outcome of a study in many subtle ways (see chapter 9). The following case involves a group of scientists' financial interest in a human cell line.

THE "MO" CELL LINE

Imagine a cancer patient's receiving treatment at a prestigious teaching hospital. The patient undergoes surgery and completes postoperative therapy. Some years later the patient discovers that members of his clinical team preserved and patented a cell line they derived from the patient. There was no disclosure, no consent, and no effort made to involve the patient in exploiting the commercial value of his unusual cells.

This was the story of John Moore, a land surveyor from Seattle, Washington. Moore was diagnosed with hairy cell leukemia in 1976 at the UCLA Medical Center. Soon after his diagnosis, Moore had his enlarged spleen removed. At the time of his surgery the prognosis for his recovery from leukemia was grim. To the surprise of his clinicians, Moore recovered after his spleen was removed, and he was thus able to return to work on the Alaska pipeline. Baffled by his recovery, clinicians at UCLA took samples of Moore's blood, bone marrow, and other body substances to determine the cause of his body's defeat of leukemia. They discovered that Moore's spleen cells produced unusually large quantities of proteins, such as interferon and interleukin, which are known to stimulate the immune system.

One of the UCLA medical scientists cultured a cell line from Moore and, in collaboration with his university, applied for a patent for the "Mo" cell line in 1981. The patent was awarded in 1984. The cell line was eventually sold to a Swiss pharmaceutical company for $15 million and brought the company several billion dollars in sales.

Moore brought suit against UCLA and the clinical researchers who appropriated his cells, charging that they failed to disclose their financial interests in his cells and that he was entitled to share in the profits from the "Mo" cell line and any other products from the research on his biological materials. The case was eventually heard by the California Supreme Court. In 1990, the court ruled against Moore's claim that he was entitled to share in the profits from the "Mo" cell line. The court's majority held that a donor does not have a property right in the tissues removed from his or her body and that to favor Moore's claim would "hinder research by restricting access to the necessary raw materials" of science thereby interfering with its progress.[9]

While Moore lost on the argument that he was entitled to commercial rights over his body parts excised during or after surgery, the court ruled favorably on another of Moore's claims. The judges found that his physicians had an obligation to disclose their financial interests in his cells. They viewed the financial disclosure as an extension of the responsibility of informed consent. The court's majority stated, "We hold that a physician who is seeking a patient's consent for a medical procedure must, in order to satisfy his fiduciary duty and to obtain the patient's informed consent, disclose personal interests to the patient's health, whether research or economic, that may affect his medical judgment."[10] The court rejected the defendant's argument that a physician who orders a procedure partly to further a research interest unrelated to the patient's health should be excused from disclosure, lest the patient might object to participation in the research. Rather, the court ruled that the reason a physician must make disclosure is that any such personal interests (in this case, the financial or research interests in the cell lines) may affect clinical judgment. A potential conflict of interest exists where any motivation for a medical procedure is unrelated to a patient's health. The majority opinion held that

> a physician who treats a patient in whom he also has a research interest has potentially conflicting loyalties. This is because medical treatment decisions are made on the basis of proportionality—weighing the benefits to the patient against the risks to the patient. . . . A physician who adds his own research interests to this balance may be tempted to order a scientifically useful procedure or test that offers marginal, or no benefits to the patient. The possibility that an interest extraneous to the patient's health has affected the physician's judgment is something that a reasonable patient would want to know in deciding whether to consent to a proposed course of treatment. It is material to the patient's decision and, thus, a prerequisite for informed consent.[11]

A state supreme court is usually not precedent setting for other states. Unless a decision is reached by the Supreme Court or unless Congress legislates this issue, other courts may read conflicts of interest differently. They may not interpret the commercial appropriation of disposable body tissue or cells as relevant to patient health and therefore a disclosable interest within the informed consent framework.

In the next case, conflict of interest is linked directly to a clinical trial.

THE DEATH OF JESSE GELSINGER

Suppose you are a patient being treated at a prestigious university medical center. You have agreed to be part of clinical trial that is not designed to improve your condition but may help others with more severe cases of the disease. You are asked to review and sign the "informed consent" forms, which outline the goals, anticipated benefits, and potential risks of the experiment. The "informed consent" process, however, does not reveal that the principal investigator of the study has a financial interest in the procedures being tested in the clinical trial. Do you have a right to that information? Would it affect your decision to enter the trial?

At age two, Jesse Gelsinger was diagnosed with a rare metabolic liver disorder called "ornithine transcarbamylase deficiency" (OTC). The disorder inhibits the body's ability to breakdown ammonia, a natural byproduct of metabolism. The body's buildup of ammonia can have lethal effects if not treated. Gelsinger's condition was treated with a low-protein diet and drugs. Once treated, it was not considered life threatening for him.

In September 1998, when Jesse Gelsinger was seventeen years old, his physician informed him about a clinical trial that was about to be started at the University of Pennsylvania Medical School to treat his disorder with a new type of gene therapy. The research protocol involved the use of a disabled adenovirus as the vector to carry the corrective genes into the cells of patients. Scientists hoped that the cells with the corrective genes would eventually replicate and encode sufficient amounts of the enzyme to metabolize ammonia. A year after he was informed about the trial, Gelsinger returned to Philadelphia to begin the therapy. He was told that the experimental gene therapy treatment was not expected to cure his form of the metabolic disorder; rather, it was designed to test a treatment for babies with a fatal form of the disease. Notwithstanding that fact, Gelsinger agreed to participate as an act of beneficence to others with whom he could empathize.

A clinical researcher drew thirty milliliters of the adenovirus containing the corrective gene and injected it into Mr. Gelsinger. That evening Gelsinger became ill. His body temperature rose to 104.5 degrees Fahrenheit. The ammonia level in his blood rose to more than ten times the normal level. He was placed on dialysis. Within days, Gelsinger's condition rapidly declined as he suffered from multiple-organ system failure. Four days after his initial treatment with the bioengineered adenovirus, the young Gelsinger deteriorated until he became brain dead. The cause of his death was attributed to his immune system's response to the adenovirus vector used in the gene therapy experiment.

During the inquiry following Gelsinger's death, it was disclosed that the director of the University of Pennsylvania's Institute for Human Gene Therapy, James Wilson, founded a biotechnology company called Genovo, Inc. Both he and the University of Pennsylvania had equity stakes in the company, which had invested in the genetically altered virus used in the gene therapy experiment.

Wilson and one of his colleagues also had patents on certain aspects of the procedure. At the time, Genovo contributed a fifth of the $25 million annual budget of the university's gene therapy institute and in return had exclusive rights over any commercial products.[12] The informed consent documents made no mention of the specific financial relationships involving the clinical investigator, the university, and the company. The eleven-page consent form Gelsinger signed had one sentence, which stated that the investigators and the university had a financial interest in a successful outcome. When Genovo was sold to a larger company, James Wilson had stock options that were reported to be worth $13.5 million; the university's stock was valued at $1.4 million.[13] According to the report in the *Washington Post,* "numerous internal Penn documents reveal that university officials had extensive discussions about the possible dangers of such financial entanglements."[14]

The Gelsinger family filed a wrongful death lawsuit against the university, which was eventually settled out of court for an undisclosed sum of money.[15] One of the plaintiff's allegations in the suit was that the clinical investigator overseeing his trial had a conflict of interest that was not adequately disclosed prior to Jesse Gelsinger's involvement. They argued that the financial interests in conjunction with other undisclosed or downplayed risks might have altered the family's risk–benefit estimate before entering the trial, which might have saved young Gelsinger's life. After the University of Pennsylvania settled with the Gelsinger family, the administration announced new restrictions on faculty who are involved in drug studies when they have equity in companies sponsoring the research.

In the aftermath of the Gelsinger case, the Department of Health and Human Services (DHHS), under the leadership of Donna Shalala, held hearings on whether the financial interests of clinical investigators should be listed on informed consent information given to prospective candidates for clinical trials. In a draft interim guidance document issued in January 2001, DHHS suggested that researchers who are involved in clinical trials disclose any financial interests they have to institutional review boards that monitor other ethical issues and possibly to the individuals who are deciding whether they plan to participate as human subjects (see chapter 12).[16] Leading scientific and medical associations, including the Federation of American Societies of Experimental Biology (FASEB) and the Association of American Medical Colleges (AAMC), opposed the idea of a guidance document for clinical trials, arguing that it overregulates medical research without contributing to the safety of patients.

In March 2003, DHHS Secretary Tommy Thompson issued a second draft guidance document titled, "Financial Relationships and Interests in Research Involving Human Subjects,"[17] which contained some weaker language than the interim document it replaced. Both drafts of the guidance document do not have the force of an agency rule as they are not designed to be regulatory, but rather provide considerations and recommendations to individual institutions, which decide for themselves whether they will require the disclosure of financial conflicts of interest to human subjects participating in clinical trials. The 2001 interim guidance document stated that a conflict that cannot be eliminated *should* be disclosed in the human subject consent document. In contrast, the 2003 draft guidance document merely recommended that investigators consider including financial disclosures in the consent document. While DHHS acknowledges that "some financial interests in research may potentially or actually affect the rights and welfare of subjects,"[18] the agency leaves the responsibility of how to protect those subjects to the individual institutions.

With millions of Americans participating in over 40,000 clinical trials in 2002, about 4,000 of which are supported by NIH, research scientists and the companies sponsoring those trials are concerned that the additional disclosure requirements that do not have a direct bearing on the safety or benefits of the trials would create unnecessary impediments to attracting human volunteers. On the other hand, the decision to become a human volunteer in a medical experiment can be one of the most important choices a person can make in his or her life. Why shouldn't a prospective volunteer know everything of relevance when determining whether to enter a relationship of trust with a clinical investigator?

SEARCHING FOR IMPARTIAL EXPERTS

Over a million women have availed themselves of breast implants in the past twenty-five years. Over time, reports began to surface that silicone, which was used extensively by implant surgeons during the 1970s and 1980s, began leaking from the implants. It was speculated that such leaking may be linked to a variety of rheumatological and immunological diseases. Thousands of suits were initiated on behalf of women who claimed they were harmed by the silicone implants. Highlighted in the media and in popular magazines was the graphic testimony of alleged victims who gave accounts of what they claimed were the results of diverse tissue and immunological effects. The federal court cases were consolidated for pretrial proceedings before Judge Sam Pointer of the United States District Court, for the Northern District of Alabama. Three of the defendants were Bristol-Myers Squibb, 3M Corporation, and Baxter International.

Judge Pointer was faced with a case that involved a voluminous body of contested scientific findings, medical claims, and expert opinions. He appointed an expert committee called the "Silicone Breast Implant National Science Panel," which consisted of four scientists in the fields of immunology, epidemiology, toxicology, and rhematology. The charge given to the panel was "responsibility of evaluating and critiquing pertinent scientific literature and studies bearing on issues of disease causation in the breast implants litigation."[19] Among the four appointed medical experts was Peter Tugwell—a physician, professor, clinical rheumatologist, and clinical epidemiologist, who also served as chair of the Department of Medicine at Ottawa Hospital in Canada. Tugwell joined the National Science Panel in April 1996.

Tugwell's institution provided incentives to faculty to attract industry funding for research and clinical trials. Documents from the institution revealed its interest in hiring someone to advance the goals of a closer linkage with commercial entities: "The person would be responsible for finding and increasing the number of clinical trials with the pharmaceutical industry to developing marketing plans for new products and services resulting from our research labs, to find commercial sponsors for successful technology transfers and to protect the intellectual property rights of our scientists."[20]

One of the contested issues in this case was whether Tugwell had a significant conflict of interest growing out of the associations he had with one or more companies who were parties of the litigation. In this case, silicone breast implants and conflicts of interest became intertwined in the litigation.

The court appointed Tugwell after reviewing his disclosures and concluding that he did not have any present or past affiliations, relationships, or associations that would impair his neutrality or objectivity.[21] Tugwell admitted he had developed relationships with parties to the suit involving arthritis but that these relationships had nothing to do with silicone breast implants and thus did not represent a conflict of interest for his participation on the expert panel. He indicated in his deposition that his curriculum vitae included a large number of industry-funded studies, and the fact that he was chosen for the panel meant that it was acceptable for him to continue those activities. He signed a statement in July 1996 that asserted, "I know of no reason why I cannot serve the Court as a neutral, unbiased and independent expert."[22] The following were some of the conflict-of-interest allegations against Tugwell.

Prior and subsequent to Tugwell's selection to the science panel, he had contact with several companies involved in the litigation. He solicited and received funding from Bristol-Myers and 3M Corporation on behalf of a professional organization called OMERACT, which runs conferences. While Tugwell was serving on the science panel, he signed a letter soliciting funds in support of such conferences in August 1997, which was sent to many companies including Bristol-Myers. This type of letter reveals the hidden agenda of medical sponsorship and influence peddling that becomes a marketing tool for academic institutions to lure industry support. The letter stated:

> We think that support for such a meeting would be very profitable for a company with a worldwide interest in drugs targeted in these fields. The impact of sponsorship will be high as the individuals invited for this workshop, being opinion leaders in their field, are influential with the regulatory agencies. Currently we are seeking major sponsors to pledge support of U.S. $5,000 and $10,000. These major sponsors will be given the opportunity to nominate participants to represent industry's interest and to participate actively in the conference.[23]

Before joining the court panel, Tugwell signed a conflict-of-interest questionnaire that disclosed his participation in the 3M funding. In August 1996, he conferred with the court about the questionnaire. Tugwell reported he had not received any personal funds from his work with OMERACT. Based on his disclosures, the decision was made by the court not to disqualify him. The 3M Corporation contributed $5,000, and Bristol-Myers Squibb contributed $500 to one of OMERACT's meetings. These contributions were made during the selection process of the court's scientific panel. Because Tugwell was not disqualified for his declared OMERACT activities when he was being considered for appointment to the panel, he believed it was ethically acceptable for him to continue his participation with the group while serving on the expert panel.

Tugwell engaged in discussions with Bristol-Myers as they explored the possibility of his involvement with the company, including his potential role as a clinical investigator and his service on a company medical safety board. In November 1998, he entered into a contract as a clinical investigator with Bristol-Myers for one of the company's products. It included an agreement that he would hold confidential any information he received from the company that was not already public.

In Tugwell's deposition, he was queried by the plaintiff's attorney about his relationship with one of the defendants in the silicone breast implant litigation:

> *Q.* So as I understand it . . . as of January 11, 1999, while serving on the science panel, you had entered into two contracts with Bristol-Myers Squibb, is that correct, the consulting contract, and the contract relating to the clinical trial?
>
> *A.* Again, this connection you're making is, in my opinion, not relevant because my involvement in the breast silicone implant litigation in no way related with any discussion I had with anyone else either in Bristol-Myers Squibb or any other company.[24]

The plaintiff's lawyers filed a motion to vacate Tugwell's appointment as a member of the National Science Panel because of his conflicts of interest. Judge

Pointer ruled that the matters cited by the plaintiff's Steering Committee have not affected or influenced Tugwell's work as a court-appointed expert; that Tugwell does not have a conflict of interest; and that he acted neutrally, objectively, and impartially.

One of the plaintiff's lawyers commented: "In today's world you cannot assume you can get neutral witnesses in circumstances involving pharmaceutical and medical devices from academia. . . . There should be no assumption you can have neutral experts in this kind of a case. Everyone should get the best experts they can and the jury will decide who is most credible and who has the best evidence."[25] The plaintiff's attorneys argued that the judge should have excluded Tugwell and the National Science Panel report, which was written collectively, because of Tugwell's association with two of the defendants.

Three members of the National Science Panel convened by Judge Pointer published a commentary in the *New England Journal of Medicine* (*NEJM*) in March 2000 in which they discussed their experiences on this unique advisory group to the federal judiciary. Tugwell was one of the authors of the commentary.[26] They wrote, "The judge overseeing all federal cases involving silicone-gel-filled breast-implants, decided that the questions raised by breast-implant litigation were of sufficient magnitude and complexity to merit the appointment of a neutral panel of scientific experts."[27] The *NEJM* has one of the most stringent conflict-of-interest policies among medical journals, but the editors saw no reason to require Tugwell to disclose that he had a conflict of interest with regard to his declaration of neutrality. Any ordinary meaning of that concept would seem to apply in this case because Tugwell had active financial relationships with parties to a litigation for which he was claiming to be a neutral party.

One of the attorneys for the plaintiffs wrote the editor in chief of *NEJM* informing her of one of the author's conflicts of interest related to the commentary. The response from a deputy editor was that *NEJM's* conflict-of-interest policy does not apply to "Sounding Board" articles, the topic of which "is often about a matter of policy or ethics, and not for example, which medication to prescribe for a particular disease."[28]

This case contrasts so vividly the differing interpretations of conflict of interest applied to medical experts in highly politicized and sensitive litigation. The judge in this situation considered the expert panel member sufficiently buffeted from conflicts of interest despite his acknowledged relationships with parties to the case, which included financial remuneration. His decision implied that the *subject* of the relationship between the court's experts and the defendants bore more weight than the *fact* of the relationship. Some would argue that individuals would be excused as jurors if they had the same relationships with the defendant, yet no one doubts the significance of bias in a court-appointed expert with conflicts of interest.

This case also highlights the role of corporations in influencing science. During discovery, documents were revealed that companies that were defendants in this case funded and helped to shape the research that was designed to demonstrate that their products were not responsible for any of the diseases alleged by the plaintiffs. After studies were proceeding forward, one company asked that the study design be changed to influence the statistical significance of the study. The influence by a sponsor of how a study should be done to meet the sponsor's interests, rather than deferring to the investigator's best judgement for answering a medically significant question, is an important and understudied source of bias.

The relationship between conflict of interest and bias has been downplayed within the scientific community to protect the new entrepreneurial ethos in academia. The next chapter explores this theme, which leads us to two relevant questions: First, are we likely to expect more bias from corporate-sponsored research? Second, if so, can it be managed successfully?

NOTES

1. I chose August 27, 2001, the day I was beginning this chapter.
2. The *Random House Dictionary* defines "conflict of interest" as "the circumstance of a public officeholder, business executive, or the like, whose personal interests might benefit from his or her official actions or influence."
3. Andrew Stark, *Conflict of Interest in American Public Life,* (Cambridge, Mass.: Harvard University Press, 2000).
4. Stark, *Conflict of Interest,* 123.
5. James Dao, "Rumsfeld Says He Is Limiting Role Because of Stock Holdings," *New York Times,* August 24, 2001.
6. D. F. Thompson, "Understanding Financial Conflicts of Interest: Sounding Board," *New England Journal of Medicine* 329 (1993): 573–576.
7. Association of American Medical Colleges, "Guidelines for Dealing with Faculty Conflicts of Commitment and Conflicts of Interest in Research," *Academic Medicine* 65 (1990): 488–496.
8. Department of Health and Human Services, National Institutes of Health, Conference on Human Subject Protection and Financial Conflicts of Interest, Plenary Presentation (Baltimore, Maryland, August 15–16, 2000) at aspe.hhs.gov/sp/coi/8-16.htm, p. 35 (accessed October 31, 2002).
9. *Moore v. Regents of the University of California,* 793 P. 2d 479, 271. Cal Rptr. (1990).

10. *Moore v. Regents of the University of California,* 793 P. 2d 479, 271. Cal Rptr. (1990).

11. *Moore v. the Regents of the University of California,* 51 Cal. 3d 120; P.2d 479.

12. Sheryl Gay Stolberg, "University Restricts Institute after Gene Therapy Death," *New York Times,* May 25, 2000, A18.

13. Jennifer Washburn, "Informed Consent," *Washington Post Magazine,* December 20, 2001, 23.

14. Washburn, "Informed Consent," 23.

15. Deborah Nelson and Rick Weiss, "Penn Researchers Sued in Gene Therapy Death," *Washington Post,* September 19, 2000, A3.

16. Eliot Marshall, "Universities Puncture Modest Regulatory Trial Balloon," *Science* 291 (March 16, 2001): 2060.

17. Department of Health and Human Services, Office of Public Health and Science, Draft, "Financial Relationships and Interests in Research Involving Human Subjects: Guidance for Human Subject Protection." *Federal Register* 68 (March 31, 2003): 15456–15460.

18. "Financial Relationships and Interests in Research Involving Human Subjects," March 31, 2003.

19. Deposition of Dr. Peter Tugwell, United States District Court, Northern District of Alabama, Southern Division, April 27, 1999. File #CV92-1000-S, p. 467.

20. Deposition of Dr. Peter Tugwell, File #CV92-1000-S, p. 447.

21. Deposition of Dr. Peter Tugwell, File #CV92-1000-S, p. 772.

22. Deposition of Dr. Peter Tugwell, File #CV92-1000-S, p. 462.

23. Deposition of Dr. Peter Tugwell, File #CV92-1000-S, p. 488.

24. Silicone Gel Breast Implant Products Liability Litigation (MDL 926). Deposition of Peter Tugwell, U.S. District Court, District of Alabama, Southern Division, April 5, 1999, p. 116.

25. Ralph Knowles (attorney), interview, November 29, 2001.

26. Barbara S. Hulka, Nancy L. Kerkvliet, and Peter Tugwell, "Experience of a Scientific Panel Formed to Advise the Federal Judiciary on Silicone Breast Implants," *New England Journal of Medicine* 342 (March 16, 2000): 912–915.

27. Hulka, Kerkvliet, and Tugwell, "Experience of a Scientific Panel," 912.

28. Robert Steinbrook, deputy editor, *New England Journal of Medicine,* letter to Ralph I. Knowles Jr., attorney, July 21, 2000.

9

A QUESTION OF BIAS

I get asked two questions most frequently and persistently by journalists about conflicts of interest. The first is: What difference does it make from whom scientists obtain their funding, to what companies they consult, or in what firms they have equity? After all, the universally held norms of scientific inquiry in pursuit of the truth make other relationships irrelevant as long as scientists are totally and uncompromisingly invested in that pursuit. A scientist who violates the canons of his or her profession simply to satisfy a sponsor, the argument continues, would soon become a pariah and lose standing in the profession. Whatever a scientist might gain, let us say in financial reward, is hardly worth the loss of professional standing.

This train of thought then leads us to the following queries: Given the incentives for following the experimental results wherever they lead, what evidence is there that the private investment in academic research plays any role in shaping the outcome of a study? Is there a funding effect in science? Would a major shift in funding from public to private sponsorship affect the quality or objectivity of the results?

The second question I frequently get asked is: Why place so much attention on scientists' personal financial interests when other types of interests—such as their predilection toward particular hypotheses, their need to satisfy a supportive funding agency, or their drive to achieve professional standing—are more significant factors in explaining behavior? No one doubts that scientists have their preferred hypotheses or theoretical frameworks for explaining the physical world. But these are part of the written record. Intellectual predilections are open to view by other scientists; financial conflicts of interest, however, remain outside the intellectual sphere. Biases associated with personal financial interests are surreptitious and therefore more insidious than the normal interests held by academic scientists. The passion to demonstrate beyond the skepticism of peers that a hypothesis is confirmed is what we expect of our scientists. The passion to make fortunes from the commercialization of knowledge does not in itself advance science, and it is likely to contaminate the pure drive to create knowledge.

No established research program exists that looks at these issues, and only a few journals are dedicated to exploring sources of bias (and even they are newcomers on the scene).[1] In particular, the design of research that examines whether there is a funding bias or a conflict of interest can be difficult and complex to carry out. But what cannot be overstated is the importance of the response to the question about whether financial conflict of interest is associated with bias. For instance, the editors of *Nature* placed a high burden of proof on ethicists, social scientists, and science policy analysts to demonstrate a link between financial interests in a research area and bias in the outcome.[2] Of course, examples of research exist that are carried out by companies that reveal a bias toward the very products they produce, but much of that research is not published in peer-reviewed journals. For the issue of bias to be seriously posed, we must consider whether it appears in the core of academic science.

SUSPECT SCIENCE

I shall begin this inquiry with a discussion of the meaning of bias in scientific research. Most generally, the term "bias" means "tending or weighted toward a particular outcome." In a research study, a factor that influences the results is said to bias the study when it is not considered a variable that plays an explicit role in the study. For instance, when investigators choose a sample that is not random or representative of the subject matter under investigation, it therefore contains a certain bias toward a particular outcome measure. If we wished to test whether smoking causes lung cancer and if we inadvertently chose a population that works in high-dust environments, we would be biasing the outcome unless we included "dust" as a variable.

Another form of research bias relates to the type of questions being asked in a study. Polling experts are well schooled in understanding that polling responses depend on the way the questions are framed. This concept is also well understood by polygraph experts. In these cases, the results of the poll might easily be misconstrued. A simple reformulation of the question could change the subject's response. For example, suppose the question was, "Do you approve of cloning of human reproductive cells?" Some respondents may not know that the cells will only be used for research and will not play any role in reproduction. Their response could be quite different if the question were: "Do you approve of cloning human reproductive cells that are used exclusively for research?"

Bias can also enter into the type of experiment one does. Suppose you wanted to study the health effects of a certain industrial chemical. The experimental design might use adult mice with specific doses of the chemical, but this chemical may break down rather quickly into a much more toxic metabolite. Moreover, the effect of the chemical's metabolite might be seen more readily in very young mice. A company that is funding a study might choose the experimental design that is least likely to manifest toxic effects. We might say that there was bias if the experimental design were consciously chosen to minimize findings of a hazard. Of course, many standard experimental protocols use adult mice, and they are not necessarily biased.

Bias can enter into the interpretation of results even when the experiments are well designed and not weighted toward a particular outcome. Different statistics are used for the interpretation of data, and some statistics are more likely to support the null hypothesis (i.e., that there is no effect). Most studies are published with discussions of the data. These discussions give scientists an opportunity to interpolate their data. The discussion area of the article is where we find differences among scientific analysts based upon subtle and not-so-obvious factors in their background or their relationships with sponsors of the research.

Bias also exists in the selection of evidence. Most published research on health effects of chemicals usually end with a plug for additional studies. One study alone is rarely sufficient to make the case because scientists must look at the weight of evidence from multiple studies. But which studies are selected? And how are they weighted? It is the scientist's role to collect the studies that build the evidence. Discretionary decisions can easily enter into the process.

A few journals have as a policy that they do not accept review articles (which are based on the scientist's assessment of the literature) or editorials (which involve normative judgments based on weight of evidence) from individuals with a conflict of interest. The editors of these journals believe that it is more difficult for readers to recognize the bias that emerges from conflicts of interest in these types of writings as compared to data-driven studies.

The adage "don't bite the hand that feeds you" has a special significance in scientific research. When government funds basic science, it does not have a vested interest in a particular outcome. Given the transparency of the funding and the peer-review process, government agencies have to be very careful of not appearing to tease out or shape scientific results that meet a political perspective, even in areas of applied research. Conservatives are more skeptical as a group than liberals or independents about the reality of global warming. But even while in control of the House and Senate, Republicans could not and would not attempt to influence the outcome of science.

Privately funded science is not transparent. There are unstated agendas. Many scientists who are funded by private companies understand what results would please the company and what results would benefit the company's bottom line. If a scientist is tethered to a company's research program, then the company is likely pleased with the outcome of the research and therefore would benefit by continuing to fund it. It is not unusual for the investigator to internalize the interests of the company. For example, a researcher might understand that it is to the company's advantage to establish a research design with a high burden of proof on the human toxicity of a compound or a low burden of proof on the efficacy of a drug. Unless these tests are standardized and monitored for quality control, sufficient opportunity exists for bias, even as the vast majority of scientists are committed to scientific truth.

PRIVATIZATION OF PRISONS

Suspicion of bias surely exists when the conflicts of interest are so patently obvious—but suspicion of bias is still not bias. Consider the case of Charles W. Thomas, a professor at the University of Florida who was a stalwart defender of the movement to privatize prisons in the United States.[3] He coauthored an article in the journal *Crime and Delinquency* in 1999 reporting on recidivism rates from three Florida juvenile institutions.[4] Two of the facilities were privately operated, and the third was a public facility. The authors selected samples of juveniles released from the two private institutions and compared them with those from a state-run facility. The study reported that fewer of the private facility youth had been rearrested. Moreover, the study found that the released juveniles from the private facility had been involved in fewer serious offenses after their release. These results support initiatives for turning our publicly run prison systems into for-profit enterprises.

The sponsor of this recidivism research was the Florida Correctional Privatization Commission, a group appointed by the governor that negotiated state contracts with private correctional facilities. Thomas was a paid consultant for the commission.

The Corrections Corporation of America (CCA) was founded in 1983 as an alternative to public prisons. At one point, the CCA offered to purchase and run the entire Tennessee prison system. Thomas became one of the fourteen members of the CCA offshoot called the Prison Reality Trust Board of Trustees.

He was also affiliated with the Private Corrections Project located at the Center for Studies in Criminology and Law at the University of Florida. The project, which was primarily industry-funded and which received over $400,000 from private prison companies, paid Thomas a summer salary of more than $25,000. The largest sponsor of the project was the CCA.

Some sociologists who reviewed the recidivism study identified a troubling bias. A one-year follow-up was used to evaluate the released inmates. Commentators viewed this time frame as too short a period from which to draw any valid conclusions about recidivism. According to one sociologist, as much as 40 percent of the recidivism by released offenders takes place six to ten years after their release.[5] Given the conflicts of interest, critics of the Thomas case study can only suspect a bias, but not prove one:

> When a man associated with the private prison industry in a money-making position uses such a short follow-up period in his study it inevitably arouses the suspicion that there was a desire, perhaps overriding, to get the favorable news legitimated by its appearance in a scholarly publication. The prior placement of the study results on the internet also can be suspected to be an attempt to show as soon as possible what are claimed to be the advantages of correctional privatization.[6]

The suspected bias is further supported by a 1996 General Accounting Office (GAO) report that analyzed studies comparing the costs of public and private correctional institutions in California, Tennessee, Washington, Texas, and New Mexico. The Tennessee studies were the most sound. For those, the GAO found virtually no difference in the average inmate costs per day between public and private facilities. The Private Corrections Project not only asserted that private prisons were more efficiently run but that they had less recidivism.[7]

Of course, few investigators would easily admit that their conflicts of interest can bias their research. Again, the appearance of conflict of interest does not prove bias. Therefore, notwithstanding the circumstantial evidence in this case, it takes definitive studies that compare the outcome of research by scientists with and without conflicts of interest. While these studies are difficult to perform and not readily funded, the few that have been published are provocative nonetheless.

CLINICAL TRIALS ON NEW VERSUS OLD THERAPIES

Published studies on funding bias can be found in the medical literature. One of the earliest I discovered appeared in 1986 in the *Journal of General Internal*

Medicine, the official journal of the Society for Research and Education in Primary Care Internal Medicine.[8] The author examined 107 controlled clinical trials designed to determine whether a new drug therapy was effective. Pharmaceutical companies typically favor new therapies over older ones. The logic is that patents for older therapies eventually reach maturity and become subject to competition; new therapies, however, usually mean new patent protection and thus larger profits. The question asked by the author was whether there was an association between the source of funding and the preference for new therapies. Out of the 107 trials, 76 favored the new therapy and 31 favored the traditional therapy. Of those papers favoring the new therapy, 43 percent were supported by drug companies whereas 57 percent were supported by nonprofit institutions (government, foundations, or university funds). Of those papers favoring traditional therapy, 13 percent were supported by drug companies whereas 87 percent came from nonprofit institutions.

This study showed that there was a statistically significant association between privately funded studies and the "favoring of new therapies." We call this effect "funding bias." In other words, private funding can bias the outcome of studies toward the interests of the sponsor. How do we explain the fact that so few of the pharmaceutically supported trials (13 percent) favored the traditional therapy? Why, in fact, would a company want to publish results indicating that a competitor's drug was better than its own product? Of course, explanations other than "funding bias" may exist to account for these results. After all, nearly three-quarters of the published studies favored the new therapy, and that support was almost evenly split between drug company sponsorship and other funding.

Drug companies may simply be hesitant to fund and publish reports that are likely to favor traditional therapies over new ones or that report negative outcomes of new drugs. Alternatively, the scientists supported by pharmaceutical companies may have had some inside knowledge of which others were not apprised that reduced their estimate of the traditional therapy. The author of the study reported, however, that "in no case was a therapeutic agent manufactured by the sponsoring company found to be inferior to an alternative product manufactured by another company."[9] That statement alone is worthy of consideration when the theory of funding bias is discussed.

Among the tens of thousands of clinical trials occurring each year, most are funded by for-profit companies seeking to gain FDA approval for new drugs, clinical procedures, or medical devices. The randomized clinical trial is the gold standard for clinical research. It divides the human subjects who meet the criteria for entering the trial into two or more random groups so that the researchers can avoid selection bias in the evaluation of a treatment or a drug. Regarding random sampling, two Danish researchers posed the following question: Does financial or other competing interests affect the interpretation of the results of randomized clinical trials? In particular, do such trials tend to favor interventions that benefit their for-profit sponsors? The investigators chose to focus their study on the internationally distinguished *British Medical Journal* (*BMJ*) because the editors of that journal require authors to declare their interests—financial or otherwise.[10] They identified 159 original randomized clinical trials published in *BMJ* from 1997 to June 2001. In ninety-four trials, the authors indicated that they had no competing interests; in sixty-five trials, the authors declared that they received funding from for-profit organizations (defined as a competing interest). The trials covered several fields of clinical medicine, including psychiatry, orthopedics, and cardiology. The study results showed that "authors' conclusions were significantly more positive toward the experimental intervention in trials funded by for-profit organizations alone compared with trials without competing interests."[11]

Studies of this type cannot tell us the cause behind the association between for-profit sources of funding and results that favor one type of therapy. It is possible that when companies fund a clinical trial, they demand more pretrial assurances of its success than nonprofit groups. As the authors of the *BMJ* study state, "Profit organizations, by skill or by chance, may fund only those trials in which the experimental intervention is significantly better than the control intervention." It is also possible that clinical investigators, through various subconscious mechanisms that act within the discretionary range afforded to scientists, weight their interpretations of data in the sponsor's favor.

While other studies also confirm a sponsor effect on experimental treatments,[12] social scientists have not been able rule out other plausible explanations beyond conflict of interest to account for the outcome. For example, another study found that there was a higher proportion of successful interventions reported among commercially funded randomized trials than those funded by government or the nonprofit sources. The authors believe it had more to do with the design of the trials, where "preferential support was given to trials that had a greater chance of favoring one intervention over another" than with the science or scientists conducting them.[13]

HEART DRUGS AND INDUSTRY SPONSORSHIP

One of the most elegant and influential studies that demonstrated an association between funding source and outcome was published in 1988 by a Canadian research team at the University of Toronto. This study appeared in the distinguished *New England Journal of Medicine* (*NEJM*), considered by many to be one of the half-dozen leading medical journals in the world.[14]

In this study, the authors began with the question, "Is there an association between authors' published positions on the safety of a drug and their financial relationships with the pharmaceutical company?" They focused their study around a class of drugs called "calcium channel blockers" (CCBs, also called "channel antagonists"), which are used to treat hypertension. Their choice was based on the fact that the medical community debated the safety of these drugs. The authors performed a natural experiment to investigate whether the existing divisions among researchers over the drug's safety could be accounted for by funding alone—that is, the study tested whether favorable outcomes for CCBs would be correlated with pharmaceutical industry funding.

First, the authors identified medical journal articles on CCBs between March 10, 1995, and September 30, 1996. Each article (and its author) was classified as being either supportive, neutral, or critical with respect to these drugs. Second, the authors were sent questionnaires that queried them over whether they received funding in the past five years from companies that either manufacture CCBs or manufacture products that compete with them. The investigators ended up with 70 articles (5 reports of original research, 32 review articles, and 33 letters to the editor). Among the seventy articles, eighty-nine authors were assigned a classification (supportive, neutral, or critical). The completed questionnaires about the authors' financial interests were received from sixty-nine authors. The study's results showed that the overwhelming number of supportive authors (96 percent) had financial relationships with manufacturers of CCBs while only 37 percent of the critical authors and 60 percent of the neutral authors had such relationships. The authors of the *NEJM* study wrote that "the results demonstrate a strong association between authors' opinions about the safety of calcium-channel antagonists and their financial relationships with pharmaceutical manufacturers."[15]

Once again, no associational study can demonstrate causality. It is possible that corporate sponsorship of studies may have had no effect on the researchers' views about CCBs. Those views might have been formed before the scientists developed a financial relationship with the companies. After all, one-third of the critical authors did have a financial tie with the pharmaceutical industry. But what else could explain the discrepancy?

The theory of funding bias does not claim that it can account for individual behavior. What it does claim is that in a population of researchers, the financial association of authors with firms will skew the outcome of results in favor of the firm's interests. The fact that a conflict of interest coincided with or preceded a biased outcome or scientific misconduct does not mean it was the cause of the bias or the motivation behind misconduct. But it does represent prima facie evidence in accounting for the outcome, especially when other explanations seem more remote. Moreover, the appearance of a conflict of interest provides circumstantial evidence of an influencing motivation.

Consider the case of a clinical researcher at the University of California, San Diego (UCSD). Dr. Maurice Buchbinder did clinical research on the Rotablader—a drill-like device used to remove plaque from arteries. Buchbinder was a large-equity holder in Heart Technologies, which manufactured the device. The FDA audited Buchbinder's research in 1993 and reached a finding that there were significant deficiencies in his clinical work, such as his failure to conduct proper follow-up and to report adverse effects. UCSD stopped Buchbinder from doing research on patients. When an investigator has strong financial stake in the outcome, violations of clinical research are viewed through a different lens, as the *New York Times* header on the story attests: "Hidden Interest: When Physicians Double as Entrepreneurs."[16]

If the "funding effect" is found in other areas where science has financial interests, then how do we explain it? Is it simply that researchers take on the interests of their private-sector sponsors? It is probably much more subtle than that. Science is a social process; thus, despite the shared norms about objectivity and truth, a population of scientists can be influenced by the values of private funders, who can especially influence interpretations that are not cut and dry. Where there is wiggle room in science, it will tend to shift toward the sponsor's interests. Clinical trials and other data driven studies on humans, while vulnerable to bias, do not provide the greatest wiggle room for multiple interpretations, certainly not nearly as much as economic studies. The next case shows how funding bias enters into cost-effectiveness studies.

COST-EFFECTIVENESS OF DRUGS

In the field of pharmaceutical development, the two pillars of drug registration are safety and efficacy. But drugs that meet these necessary criteria might still not find a home in the pharmaceutical marketplace if they fail to be cost-effective. In this period of managed care, drugs that cost far in excess of the benefits that they offer would not likely be approved for payment by prepaid health insurers, especially if other less costly (albeit less effective) therapies are available.

The cost-benefit analysis of new drugs is a field called "pharmacoeconomics." This applied subfield within economics has taken on a new role in the cost-saving health care economy. Cost-benefit studies of new drugs can determine the success or failure of a product that has been shown to be safe and effective. Just how vulnerable are these studies to "funding bias"?

A team of researchers tested the hypothesis that an association exists between pharmaceutical-industry sponsorship and the positive economic assessment of oncology drugs.[17] The research group focused its investigation on oncology drugs that were cited in the medical databases during the ten-year period between 1988 and 1998. They found six oncology drugs discussed in forty-four articles on cost-effectiveness that qualified for the study. Each of the articles was rated according to whether, on cost-effectiveness grounds, it considered the drug favorable, neutral, or unfavorable. As in other studies, the funding sources were investigated after recording the study's qualitative conclusions. Articles were classified as either pharmaceutical-company sponsored or nonprofit-sponsored. Of the forty-four articles, twenty were funded by pharmaceutical companies, and twenty-four were funded by nonprofit organizations. Unfavorable conclusions about the cost-effectiveness of drugs were reached by 38 percent of the nonprofit sponsored studies but by only 5 percent of the pharmaceutical-company sponsored studies. The authors concluded, "Studies funded by pharmaceutical companies were nearly 8 times less likely to reach unfavorable qualitative conclusions than nonprofit-funded studies and 1.4 times more likely to reach favorable qualitative conclusions."[18]

It is conceivable that there are explanations for this result, other than the systematic bias of the privately funded study. Companies may fund academic studies of drug efficacy only after they have strong evidence that the drugs work. The skewing of the results toward favorable conclusions could be the result of many company-funded studies that are not published. Nevertheless, the authors of the study assert that when researchers receive funding and honoraria from a company, these factors "may result in some unconscious bias (perhaps when qualitatively interpreting results) that influence the study's conclusions"[19] and that "the pharmaceutical companies can collaborate directly with investigators in devising protocols for economic analysis and indirectly shape the economic evaluation criteria."[20]

Pharmacoeconomic analysis does not possess a set of canonical methods that are shared by all professionals in the field—unlike, let us say, the discipline of structural engineering, which most certainly does. Many discretionary, a priori assumptions can enter into such an analysis. The plasticity of the methods makes them more vulnerable to influence by a sponsoring company. As a result, "the differences observed between studies funded by industry and nonprofit organizations may be the result of methods chosen, prescreening, or bias due to the source of funding."[21]

JOURNAL SUPPLEMENTS

Pharmaceutical companies frequently underwrite supplement volumes published by highly respected journals. The supplements contribute funds to the journal, give some scientists an opportunity to publish under the journal's logo, and allow companies to support symposia to advance their commercial interests. But the standards for review for such published articles can be very different than the standards used by the parent journal. For one thing, many of the articles published in journal supplements are not peer reviewed. Articles may be reviewed by an editor, but they are not sent out for blind review by experts in the subject matter of the article.

The articles published in journal supplements often appear in databases like MEDLINE and are housed on the shelves of medical libraries next to the parent journals. Readers of the articles may not be aware that these articles have not met the most stringent review criteria. The supplements to prestigious journals benefit pharmaceutical companies because they contain articles related to their products. But what about the potential for bias or diminished quality in publications funded by corporate sponsors?

A study was undertaken to evaluate the quality of articles in journal supplements by comparing them with the quality of articles published in the parental journal during the same period.[22] The authors of the study identified articles listed in MEDLINE from January 1990 to November 1992 on randomized trials of drug therapies published in three leading journals known to publish supplements. Through the use of a quality-assessment scoring system, the investigators gave numerical quality points for each of the 242 articles in the study sample. Their results showed that the articles they reviewed and scored on randomized control trials of drug therapies in journal supplements were generally of inferior quality compared with the articles published in the parental journals. The authors wrote, "Articles published in journal supplements provided less information to insure that adequate steps had been taken to assess compliance with the study medication."[23]

The authors also found a greater discrepancy between the number of patients randomized (the starting group) and the number of patients analyzed (the final group) in articles published in journal supplements compared with those published in the peer-reviewed parent journals. The difference between the groups is reflected in the withdrawal rate. A high withdrawal can influence the outcome and statistical soundness of a study.

According to the authors of this study, "The funding source may influence whether or not a particular study is published and thus lead to a form of publication bias. Manufacturer-sponsored publications may tend to favor the manufacturer-sponsored drugs."[24] Other studies have confirmed this effect and found a relationship between manufacturer sponsorship and favorable findings of drug efficacy and toxicity.[25]

Pharmaceutical companies are in business to get products to market. To achieve this objective, they must have a sustained relationship with academic scientists who generate credible data that are submitted to the Food and Drug Administration. They are also subject to enormous litigation settlements when a drug is found to cause death or severe illness. Other corporate sectors seek out academic scientists to protect themselves from litigious workers or consumers who file high-profile lawsuits. In these situations, companies who are being sued have much to gain from developing mutually beneficial ties with research scientists, especially when the goals and temperament of the scientist are compatible with the values and interests of the company. The investments that companies make in building bridges to academic science are like insurance policies. During litigation that is sometimes referred to as "toxic torts," where a company is charged under product-liability law, university scientists are often called upon to serve as expert witnesses on behalf of corporate defendants.

TOXIC TORTS AND ACADEMIC SCIENCE

For over a decade, CSX Transportation, the largest railroad in the eastern United States, has been dealing with worker claims that they have been brain damaged from exposure to solvents like 1,1,1 trichloroethane; trichloroethylene; and perchloroethylene.[26] CSX has made confidential settlements with certain workers, but the cases involved no admission that these and other chemicals caused the workers' illnesses.

CSX financed a research study to evaluate claims made by its workers who had been diagnosed as having chronic toxic encephalopathy. The company had contracted a professor of neurology at the University of Michigan (UM) who had served the company as an expert witness in litigation. In that capacity, the professor had access to the workers' health data. Naturally, the company financed the study. Approximately $30,000 for the professor's work, however, came from the DOW Chemical Corporation, which was the manufacturer of some of the solvents.

The UM professor had previously testified that he had never seen a patient with cognitive or behavioral problems caused by exposure to solvents in the workplace, despite the fact that other qualified professionals in medicine and neuropsychology had diagnosed CSX workers with cognitive difficulties. In the

CSX-financed study, the UM investigators were able to get a waiver from the university's Institutional Review Board (IRB) for obtaining the informed consent of the workers. This decision meant that the investigators could use any medical information they had, including information they had obtained from examinations while they served as expert witnesses on behalf of the company. Without the consent of the workers, the investigators undertook a retrospective study based on medical information obtained as part of a lawsuit. The company could then cite the results of the study, published in the *Journal of Occupational and Environmental Medicine*,[27] as evidence that the workers' claims were not grounded on science.

This case is emblematic of the way chemical corporations face potential lawsuits. They seek to fund academic scientists who are paid lucratively as expert witnesses and then as research scientists. Once the scientist signs on as an expert witness, he or she has begun to internalize the interests of the company. The company then trusts that the scientist will produce research that supports his or her testimony. It is extremely rare to find industry expert witnesses who reverse their views about the risks to human health from toxic substances after their company-funded research is completed.

But can companies that wish to influence the research agenda on toxic substances affect research that is funded from government sources? This issue is much more complicated because the government operates through a system of scientific peer review by which government scientist/managers disperse research funds. Ironically, a new kind of partnership between government and industry gives to the latter some influence in the decisions of what federally sponsored research to fund, a decision that has traditionally been the exclusive province of government.

In July 2001, the National Institute of Environmental Health Sciences (NIEHS), the country's premier federal agency for the study of the effects of toxic substances on human and environmental health, signed a memorandum of agreement with the American Chemistry Council (ACC), a chemical industry trade association. The goal of the agreement was to work together for improving the testing of chemicals for any human developmental and reproductive effects. According to the agreement, the ACC will contribute $1 million and the NIEHS $3 million toward a $4 million fund for research "on the mechanisms of action of potential developmental toxicants using state-of-the-art tools, including genomics and genetic animal models."[28] Representatives from the ACC serve with NIEHS scientists on a panel that screens grant applications before they are sent out for peer review. Through the ACC, the chemical industry will have privileged access to application data and proposals of independent scientists who must submit a statement that they agree to share their data with the industry representatives.

It is quite common for scientists to include preliminary findings in their proposals submitted to funding agencies. In this case, this information is not secure from being passed on to company lawyers and toxicologists before the results are published. For companies, having this access could mean getting a head start on the development of a combative approach to research findings. Moreover, scientists who are uncomfortable about having their work reviewed by representatives of the American chemical industry would exclude themselves from applying for this pool of publicly funded research.

How can industry funding of a government research program not influence the type of research that gets done? Public research of environmental health has largely been insulated from the private sector. This precedent of the NIEHS agreement with the chemical industry could be the first step toward privatizing health research. Of course, industry has valuable expertise and financial resources. But when has that expertise ever been used to withdraw a toxic chemical from the environment? The NIEHS has been the leading federal agency on issues of endocrine-disruptor research, while industry has stonewalled the question that some chemicals in low doses can disrupt the endocrine systems of humans and wildlife. The shaping of that research agenda can determine whether suspect chemicals such as phthalates or bisphenol-A will ever be removed from human exposure.

In *Toxic Deception*, authors Fagin and Lavelle analyzed studies of four chemicals—alachlor, atrazine, formaldehyde, and perchloroethylene—that have been targets of concern for their health effects. They checked the National Library of Medicine's MEDLINE database for articles published between 1989 and 1995 on the health effects of these chemicals—articles that were financed by corporations or industry-sponsored organizations. Of the forty-three studies they found, six were unfavorable on health effects (14 percent); five had mixed or ambivalent findings; and thirty-two indicated that the chemicals were not harmful (favorable findings).[29]

They then surveyed the articles published during the same period that were not supported by industry. They found 118 studies, of which all but two had some discernable stake in the outcome (an insurance fund of a textile union). In this case, about 60 percent of the studies (71) found the chemicals harmful to humans (unfavorable), while the remaining were divided between studies with favorable results (27) and those that were ambivalent or difficult to characterize (20). These patterns of sponsorship of research seem to hold whether we are dealing with industrial chemicals, tobacco, or clinical drugs. The source of the funding makes a difference in the outcome in a large sample of studies.

TWEAKING THE PROTOCOLS

A "piper" is a person who plays a flutelike or wind instrument, such as a bagpipe. The popular expression "who pays the piper," in one literal context means that the music one hears is determined by who sponsors the musician. Sometimes the expression has been applied to the funding of science. Will the results of a study depend upon the source of the funding? How can science, the most reliable system we have for fixing belief, be subject to such a subjective folk principle as "paying the piper"? The answer lies in the subtle areas of study design in science.

The parents of a son born with a birth defect sued a pharmaceutical company on behalf of their child, alleging that the abnormality he was born with was caused by a drug his mother had taken during pregnancy. This story is not an unusual one. Many similar allegations and suits have been made against pharmaceutical companies over the years. This particular case was over a drug called "Bendectin," which was prescribed for morning sickness. Like many other cases, scientific and medical experts were hired by the plaintiff and defense attorneys to argue their perspectives over the plausibility that Bendectin was responsible for the birth defect, which in this case was clubfeet. The jury in the trial court found that the pharmaceutical company acted fraudulently and negligently in failing to provide proper warnings, and they awarded $19.2 million for the plaintiff.

The case was eventually heard by the Pennsylvania Supreme Court. Much of the Supreme Court's decision dealt with the arcane issues of the admissibility of expert testimony. Reversing the verdict in the trial court, the Supreme Court majority ruled that the expert testimony for the plaintiff—who argued a causal link between the mother's ingestion of the drug and the birth defect—was flawed, unreliable, and inadmissable under the current standards for scientific evidence in liability cases.

What caught my attention in this case was a dissenting opinion of one of the Supreme Court judges. The judge observed that the accepted methodology for studying Bendectin's effects on the fetus was supported mainly by the drug company. He wrote that the pharmaceutical company "largely created the 'generally accepted orthodoxy' that would freeze out viewpoints contrary to their litigation interests" and thus "subsidized or otherwise influenced most of the studies that concluded that Bendectin does not cause birth defects."[30] The current rules of evidence in federal tort cases can exclude expert testimony by scientists whose views depart from the orthodoxy. The dissenting judge summed up his points as follows: "There is something not a little offensive about an entity, creating a biased litigation-driven scientific 'orthodoxy,' and then being permitted to silence any qualified expert holding a dissenting view on the grounds of 'unorthodoxy.'"[31] To what extent does this observation by the Pennsylvania Supreme Court justice provide a clue to the way that corporations fund university science to advance their interests?

Corporations fund academic science in two general areas: research and development; and safety and toxicity studies. In the latter category, corporations have a clear interest in specific outcomes—namely, to protect themselves from the costs of unfavorable jury verdicts against their product. In those instances, some corporations invest in academia with a specific agenda in mind: to prove the null hypothesis that no causal link can be found between their product and the alleged harm claimed by a consumer (e.g., disease, abnormality, dangerous malfunction).

When corporations fund university scientists to undertake health and safety studies, it is important to know if they participate with the academic investigators in designing the study. Do they make modifications in the study's design (called "protocols") in midstream? Do they attempt to control whether the results of studies they fund are published?

The answer is that examples of all of these corporate behaviors can be found in the historical record. Because the contracts that corporations sign with university researchers are often kept confidential, the prevalence of these behaviors is not known. When cases are litigated (and thus not settled out of court), the trial discovery process reaches into the company files and often reveals the role that sponsors of research have in the study design.

The silicone-breast implant case is illustrative of how some corporations interact with academic scientists when they are funding research to defend a product that is suspected of harming consumers.[32] In this instance, the manufacturers developed a litigation strategy in response to a large number of lawsuits. According to documents introduced into the litigation, one company funded research that was designed to maximize the probability that silicone implants would not be *found* to cause a disease. The company established four conditions before it would award funds for the research. These conditions were viewed by the plaintiff's lawyers as biasing the outcome of studies against the claims of their clients.

The conditions are as follows: First, studies should look at traditional connective tissue diseases and not the atypical symptoms reported by clinicians in the literature. Second, studies should include saline as well as silicone implants. Assuming the saline implants were not a problem, they would dilute the cases of concern, reducing the possibility of obtaining a statistically significant finding that silicone caused disease. Third, the studies should use a test of significance (two-tailed) that considered both the positive and negative impacts of having silicone-breast implants, even though there were no hypotheses that silicone implants improved women's health. Fourth, all women who exhibited symptoms after 1991 should be excluded from the study. This exclusion kept the mean of "years with implant" to between seven and nine, although some experts believe it can take ten or more years for symptoms to develop.

Documents from the silicone–breast cancer litigation revealed that Dow Corning Corporation scientists persuaded Johns Hopkins University medical researchers to revise their protocols for a case-controlled study it was considering to fund.[33] There is no way of knowing from the documentation whether the changes in that protocol suggested by the sponsoring company weighed more heavily in favor of its interests. But if these protocol negotiations between investigator and sponsor were the common practice in the safety and health assessment of products, it stands to reason that, in the long run, a biasing function favoring sponsors' interests is a likely outcome. Each external study that Dow Corning funded was reviewed by its legal counsel to determine its impact on the litigation.[34]

There is also documentation in the litigation file that a researcher at the Mayo Clinic, who had access to nearly one thousand patients who had silicone implants, was approached by Dow to fund a study on the condition that the company would exercise control over "whether and what was published."[35]

When corporations are faced with potentially large settlements or jury verdicts, some will dig their heals in and undertake a campaign to construct a scientific consensus clearing it of liability. When Dow Corning was faced with over 100,000 women who had silicone breast implants, writers Stauber and Rampton reported that its public relations firm advised the company: "We must begin by identifying supportive science, scientists, across the spectrum of uses for silicone; training and supporting to get our message out . . . using them proactively to brief the trade, general and business media . . . using them reactively as a 'truth squad' to refute antagonists."[36]

The bias infused into protocol tweaking does not explicitly breach the ethics of science. As long as scientists state their methods clearly; are faithful in how they carry out the protocols of a study; and are scrupulous in how they collect and interpret their data, they are fulfilling their professional responsibility. But from the public-interest standpoint, a scientist's willingness to alter the protocols of a study in response to a private sponsor's request so that the modified design favors the null hypothesis does raise serious questions about whose interests are served by the so-called free and independent academic scientist. Only the piper knows.

MIT Professor Nicholas Ashford, in a letter published in the *American Journal of Industrial Medicine*, outlined the subtlety of bias that is introduced into scientific research. His analysis, more than any single study, demonstrates why total openness of research funding (and, in some cases, complete exclusion of conflict of interest) is vital to the public trust of science. Ashford writes:

> The avoidance of the *appearance* of conflict is every bit as important as conflicts themselves if science its to regain its proper stature in public policy debates. It is an illusion to insist that values do not shape the choices of problems addressed, data relied upon and interpreted, methodologies employed in discovery and analysis, presentations and reporting of results, and acknowledgment of contrary views and data. One does not have to manipulate data or use invalid methods tantamount to fraud to bias a scientific paper. The omission of the citation of contrary data and studies is very difficult to pick up in the process of peer review. There is considerable leeway within *acceptable* choices to investigate, interpret, and present data—and cite other studies. . . . Intentional bias in choices of methodology, data, and styles of interpretation within well-accepted limits are well-nigh impossible to detect or prove.[37]

Marcia Angel, former editor of *The New England Journal of Medicine*, commented that it was her impression "papers submitted by authors with financial conflicts of interest were far more likely to be biased in both design and interpretation."[38]

Angel's impression was validated by findings that appeared in the *Journal of the American Medical Association* from a systematic study of published research (called a meta-analysis) on the "extent, impact, and management of financial conflicts of interest in biomedical research."[39] Beginning with a screening of 1,664 original research articles, the authors culled 144 that were potentially eligible for their analysis and ended up with 37 studies that met their criteria. One of the questions the authors pursued in their study was whether there was a funding effect in biomedical research. Eleven of the studies they reviewed found that industry-sponsored research yielded pro-industry outcomes. The authors concluded:

> Although only 37 articles met [our] inclusion criteria, evidence suggests that the financial ties that intertwine industry, investigators, and academic institutions can influence the research process. Strong and consistent evidence shows that industry-sponsored research tends to draw pro-industry conclusions. By combining data from articles examining 1140 studies, we found that industry-sponsored studies were significantly more likely to reach conclusions that were favorable to the sponsor than were nonindustry studies.[40]

How do the scientific journals deal with industry-sponsored research where anything from subtle bias to outright manipulation has been documented? As we shall see in the next chapter, one group of medical journal editors opposed contractual agreements between companies and academic scientists that deny an investigator the right to examine data independently and to control the publication of a study. The issue of who influences the protocols and how much disclosure there is in privately sponsored work continues to be a subject of debate in scientific and medical ethics.

NOTES

1. These journals include *Accountability in Research: Policies and Quality Assurance*; *Science and Engineering Ethics*.
2. Editorial, "Avoid Financial 'Correctness,'" *Nature* 385 (February 6, 1997): 469.
3. This case was derived largely from Gilbert Geiss, Alan Mobley, and David Schichor, "Private Prisons, Criminological Research, and Conflict of Interest: A Case Study," *Crime & Delinquency* 45 (1999): 372–388.
4. Lonn Lanza-Kaduce, Karen E. Parker, and Charles W. Thomas, "A Comparative Recidivism Analysis of Releases from Private and Public Prisons," *Crime & Delinquency* 45 (1999): 28–47.
5. Samuel J. Brakel, "Prison Management, Private Enterprise Style: The Inmates Evaluation," *New England Journal of Criminal and Civil Commitment* 14 (1988): 174–214.
6. Geiss, Mobley, Schichor, "Private Prisons," 374.
7. U.S. General Accounting Office, *Private and Public Prisons: Studies Comparing Operational Costs and/or Quality of Service,* report to the Subcommittee on Crime, Committee on the Judiciary, House of Representatives, August 1996, GAO/GGD-96-158.
8. Richard A. Davidson, "Source of Funding and Outcome of Clinical Trials," *Journal of General Internal Medicine* 1 (May–June 1986): 155–158.
9. Davidson, "Source of Funding," 156.
10. Lise L. Kjaergard and Bodil Als-Nielsen, "Association between Competing Interests and Authors' Conclusions: Epidemiological Study of Randomized Clinical Trials Published in *BMJ*," *British Medical Journal* 325 (August 3, 2002): 249–252.
11. Kjaergard and Als-Nielsen, "Association between Competing Interests," 249.
12. K. Wahlbeck and C. Adams, "Beyond Conflict of Interest: Sponsored Drug Trials Show More Favourable Outcomes," *British Medical Journal* 318 (1999): 465.
13. Benjamin Djulbegovic, Mensura Lacevic, Alan Cantor et al., "The Uncertainty Principle and Industry-Sponsored Research," *The Lancet* 356 (August 19, 2000): 635–638.
14. H. T. Stelfox, G. Chua, G. K. O'Rourke, A. S. Detsky, "Conflict of Interest in the Debate over Calcium-Channel Antagonists," *New England Journal of Medicine* 338 (January 8, 1998): 101–106.
15. Stelfox et al., "Conflict of Interest," 101–106.
16. Kurt Eichenwald and Gina Kolata, "Hidden Interest: When Physicians Double As Entrepreneurs," *New York Times,* November 30, 1999, A1, C16.

17. Mark Friedberg, Bernard Saffron, Tammy J. Stinson et al., "Evaluation of Conflict of Interest in Economic Analyses of New Drugs Used in Oncology," *Journal of the American Medical Association* 282 (October 20, 1999): 1453–1457.

18. Friedberg et al., "Evaluation of Conflict of Interest," 1455.

19. Friedberg et al., "Evaluation of Conflict of Interest," 1456.

20. Friedberg et al., "Evaluation of Conflict of Interest," 1456.

21. Sheldon Krimsky, "Conflict of Interest and Cost-Effectiveness Analysis," *Journal of the American Medical Association* 282 (October 20, 1999): 1474–1475.

22. Paula A. Rochon, Jerry H. Gurwitz, C. Mark Cheung et al., "Evaluating the Quality of Articles Published in Journal Supplements Compared with the Quality of Those Published in the Parent Journal," *Journal of the American Medical Association* 272 (July 13, 1994): 108–113.

23. Rochon et al., "Evaluating the Quality of Articles," 111.

24. Rochon et al., "Evaluating the Quality of Articles," 112.

25. P. A. Rochon, J. H. Gurwitz, R. W. Simms et al., "A Study of Manufacturer Supported Trials of Nonsteroidal Anti-inflammatory Drugs in the Treatment of Arthritis," *Archives of Internal Medicine* 154 (1994): 157–163.

26. I used two sources for this case. Katherine Uraneck, "Scientists Court New Ethics Distinctions: Questions about Litigation and Human Research Puzzle Ethicists," *The Scientist* 15 (July 23, 2001): 32. James Bruggers, "Brain Damage Blamed on Solvent Use. Railworkers Suffer after Decades of Exposure; CSX Denies Link," *The Courier-Journal*, May 13, 2001.

27. J. W. Albers, J. J. Wald, D. H. Garabrant et al., "Neurologic Evaluation of Workers Previously Diagnosed with Solvent-Induced Toxic Encephalopathy," *Journal of Occupational and Environmental Medicine* 42 (April 2000): 410–423.

28. See ourstolenfuture.org/policy/2001-07niehs.acc.htm. Also, niehs.nih.gov.

29. Dan Fagin and Marianne Lavelle, *Toxic Deception* (Secaucus, N.J.: Carol Publishing Group, 1996), 51.

30. *Blum v. Merrell Dow Pharmaceuticals, Inc.,* Supreme Court of Pennsylvania, 564Pa.3, 764A.2d.1, decided December 22, 2000. Dissent of Justice Ronald D. Castille.

31. *Blum v. Merrell Dow Pharmaceuticals,* p. 13.

32. United States District Court, Northern District of Alabama, Southern Division. In re: Silicone gel breast implants products liability litigation, (MDL 926), Case No. CV 92-P-10000-S.

33. Letter from M. C. Hochberg, Associate Professor of Medicine, Johns Hopkins University to R. R. LeVier, Dow Corning Corp., February 21, 1991: "The protocol has been revised in response to the thoughtful critique provided by scientists at Dow Corning Corporation."

34. James R. Jenkins, Dow Corning Corporation, vice president, affidavit, July 10, 1995, record no. 0486.

35. Letter from Allan Fudim, attorney (Lester, Schwab, Katz & Dwyer) to F. C. Woodside, Dinsmore & Shohl; G. G. Thiese, Dow Corning Corp.; R. Cook, Dow Corning Corporation, July 2, 1992.

36. John C. Stauber and Sheldon Rampton, "Science under Pressure: Dow-Funded Studies Say 'No Problem,'" *PR Watch Archives* 1, no. 1 (1996), at private.org/prwissues/1996Q1/silicone11.html (accessed November 29, 2001).

37. Nicholas Ashford, "Disclosure of Interest: A Time for Clarity," *American Journal of Industrial Medicine* 28 (1995): 611–612.

38. Department of Health and Human Services, National Institutes of Health, Conference on Human Subject Protection and Financial Conflicts of Interest, Bethesda, MD, August 15–16, 2000, plenary presentation, at aspe.hhs.gov/sp/coi/8-16.htm (p. 36; accessed October 31, 2002).

39. Justin E. Bekelman, Yan Li, and Cary P. Gross, "Scope and Impact of Financial Conflicts of Interest in Biomedical Research: A Systematic Review," *Journal of the American Medical Association* 289 (January 22/29, 2003): 454–465, p. 454.

40. Bekelman, Li, and Gross, "Scope and Impact of Financial Conflicts," p. 463.

10

THE SCIENTIFIC JOURNALS

The record of achievement in science can be found in the accumulated writings of scientific publications. In particular, the journals of each discipline provide the gatekeeping function for what is considered certifiable knowledge within that discipline.

A hundred years ago, about one thousand science journals were published worldwide. Today the number of active science journals published in all languages is about thirty-five thousand.[1] Approximately seven thousand of these are published in the United States.

Within any field of science, including medical research, a hierarchy of prestige exists among journals. Some journals obtain their prestige by being published through the auspices of professional societies, such as the American Chemical Society or the American Medical Association. Most professional associations have a "journal of record" that has earned its reputation as the standard bearer for the discipline or professional association. Other scientific journals achieve prestige by the eminence of their editor in chief or by a distinguished advisory board. The stature of these journals is determined by rankings of leaders in the field. The longevity of the journal's publication and its role in the historical development of a discipline also contribute to its prestige.

RANKING OF SCIENTIFIC JOURNALS

Because science is not centralized, journals have different standards for accepting articles for publication as well as different ethical codes for authors. Some general criteria are used to judge the stature of journals. First and foremost among the criteria is whether the journal operates through a system of peer review. This process is the one by which any article submitted for publication is sent out to scientists with established expertise in the topic of the submission for critical evaluation of its methods, findings, interpretation, and significance. The acceptance rate among peer-reviewed journals varies according to the prestige of the journal; however, peer-reviewed journals, when compared to their non-peer-reviewed counterparts, are in general more difficult to get published in, more likely to find and correct errors in a study, and more likely to have higher credibility among scientists. Scientists who are being considered for an appointment, tenure, or promotion at a university (as well as some government and industry research positions) are judged primarily by their peer-reviewed publications.

Another standard used for assessing the standing of scientific journals is a quantitative measurement called the "citation index." Journals whose articles are more frequently cited in other published works are considered to be more prestigious by the scientific community. There is some circularity to this: scientists read and cite articles from sources that they believe are the most prestigious journals; those journals of prestige therefore have the highest citation rates; journals with the highest citation rates then become indicators of prestige. Science is a social institution, and as such it stands to reason that it will construct its levels of confidence in selected sources of published information. The popular culture makes similar determinations among newspapers and magazines.

An annual publication titled *Journal Citation Reports* (*JCR*), published by the Institute of Scientific Information, tabulates all the citations in a preselected list of over five thousand scientific and medical journals. The *JCR* also records citations in social science and humanities journals. Here is how it works: Each journal article contains reference citations at the end of the article. These citations are compiled on a master list that yields information on how often an article is cited and how many times specific journals are cited in references. For each publication, the *JCR* generates several indicators. Two of special interest are the "journal impact factor" and the "times cited factor." The "journal impact factor" is the average number of times articles published in a journal for a specific year were cited in the *JCR* database. "Impact factors" for journals can be compared. A high-impact factor means that the articles in that journal are cited in other publications, on average, more frequently than articles in journals with a lower "impact factor." Thus, the impact factor, which is tabulated annually for journals, provides a measure of relative standing with regards to the frequency that their published articles are cited by the scientific community.

A second indicator compiled by the *JCR* is called the "times cited factor." This index provides the total number of times an article has been cited. It is recorded annually and compiled cumulatively. Journals are indexed according to the total number of citations generated by their published articles. Those journals that publish numerous articles, even though each article is cited modestly, can have a relatively high-aggregate "times cited factor." Similarly, journals that publish few but extensively cited articles can also have a relatively high-aggregate "times cited factor."

Another consideration in journal rankings is the rejection rate of submitted articles. Journals with a high-rejection rates are considered to have more rigorous peer review. Because they are more competitive, editors can elevate the acceptance criteria for submitted papers. Even when the articles are scientifically sound, they may be rejected by the more competitive journals if they do not meet other criteria, such as their perceived importance to the field. Also, journals that have a longer duration between submission and publication may have greater quality-control measures over articles. The time delay to publication may be an indication that authors have to respond more thoroughly to reviewers' comments and that a revised article is sent out for a second review.

Finally, journals are also rated informally by their impact on the mass media. Publications like *Science, Nature, Journal of the American Medical Association* (*JAMA*), and *New England Journal of Medicine* (*NEJM*) have high visibility among science journalists. These journals reach the media first, sometimes with embargoes placed on new breakthroughs, thus heightening the anticipation of science journalists to report findings. Publications with high-media status also have a high-impact factor. The reverse, however, is not true. Some journals that are cited widely among scientists do not necessarily have a high-media impact.

Published research in preeminent scientific journals is the basis for many critical policy, regulatory, public health, and medical decisions. Journal publications also affect decisions by security investors and venture capitalists, most notably on new drugs and medical devices. Journal editors take very seriously the allegations of misconduct or ethical breaches that undermine confidence in the peer-review process.

AUTHOR CONFLICT OF INTEREST

In the last quarter of the twentieth century, U.S. academic science and medicine had become more closely tethered to commercial interests than ever before. Biologists began scrambling to exploit the new-found applications of genetic engineering. They were recruited en masse to serve on scientific advisory boards of nascent companies built on venture capital and the promise of new drugs and therapies. Ten years after passage of the Bayh–Dole Act, it is estimated that

1,000 university-industry research centers (UIRCs) had been established at over 200 universities,[2] more than doubling university-industry partnerships in one decade. Many scientists partnered with business school graduates, started their own companies, and retained their academic appointments. Several decades later, social scientist Dorothy Nelkin commented: "Science is a big business, a costly enterprise commonly funded by corporations and driven by the logic of the market. Entrepreneurial values, economic interests, and the promise of profits are shaping the scientific ethos."[3]

Editors of medical journals were first to warn society about the effects of the burgeoning conflicts of interests on medical research. In a 1984 editorial describing the changes occurring in academic medicine, Arnold Relman, then editor in chief of the *New England Journal of Medicine,* wrote that "it is not only possible for medical investigators to have their research subsidized by business whose products they are studying, or act as paid consultants for them, but they are sometimes also principals in those businesses or hold equity interests in them."[4] Based on his personal observations, Relman noted that "entrepreneurism is rampant in medicine today."[5] Soon after the editorial was published, *NEJM* introduced the journal's first policy on conflicts of interest. It became the first major medical journal to require authors of original research articles to disclose any financial ties they had with companies that made products cited in their work.[6]

Most of the observations about the scale of entrepreneurial science during the 1980s were anecdotal. In the late 1980s, however, I began studying the degree to which academic scientists in pure and applied genetics were actively pursuing commercial interests or developing formal relationships with for-profit companies. Working closely with James Ennis, a colleague of mine in sociology who specializes in social networks, I developed a national database of scientists who were serving on scientific advisory committees of new biotechnology companies. I was interested in the extent to which the culture of academic science was changing. If these changes were happening slowly, then universities would therefore have an opportunity to develop ethical guidelines and management procedures for conflicts of interest. We defined the "penetration index" as the percent of a university's biomedical faculty who were involved in the commercialization of science. We referred to that group as "dual-affiliated scientists."

The penetration index was highest at some of the nation's most prestigious universities. At MIT, Stanford, and Harvard, the percentage of dual-affiliated faculty to total faculty in the biomedical field was 31, 20, and 19, respectively, for data gathered from 1985 through 1988.[7] At Harvard, sixty-nine biomedical faculty members were affiliated with forty-three different firms; at MIT, thirty-five faculty members had ties to twenty-seven firms; at Stanford, forty faculty members had ties to twenty-five firms. The protection of trade secrets among academic scholars whose labs were across the hall from one another was becoming a common practice.[8] Changes in the behavior of academic scientists were occurring at such a rapid pace that universities had difficulty adjusting.

The results of our study demonstrated that the leading research universities were setting the example for the country. Academic scientists were not only expected to do fundamental research but to exploit the commercial applications of that research. Some observers were beginning to question whether the new trend in science would affect the quality or integrity of research. Would the commingling of commercial and pure research interests tarnish the public image of science as the disinterested pursuit of truth and objective knowledge? I began to turn my attention to scientific publications. Beginning with a thought experiment, I imagined selecting an article at random from a leading scientific journal and asking myself, "What is the likelihood that one of more authors had a financial interest in the subject matter of the work?" In collaboration with L. S. Rothenberg (a colleague at UCLA), I conducted a study to answer the question, using nearly eight hundred scientific articles published in fourteen preeminent journals. We found a 34 percent likelihood that lead authors had a financial interest in the subject matter of their publication. (See chapter 7 for more details about this study.)

In addition to our study, other indicators existed that scientific authors were more than ever cashing in on their expertise as consultants, holders of equity in biomedical companies, and patent holders. The scientific and medical journals had, for the longest period, simply avoided the responsibility to address the problems of conflict of interest in their publications.

Some journals were quite defiant in not requiring authors to disclose their interests. For example, in early 1997, the international journal *Nature* published an editorial under the title "Avoid Financial 'Correctness'," which responded to our findings about the high frequency of financial interests of academic authors in published journal articles. The editor wrote: "It comes as no surprise to find . . . that about one-third of a group of life scientists working in the biotechnology rich state of Massachusetts had financial interests in work they published in academic journals in 1992."[9] The editorial went on to assert that it had never required the declaration of interests by authors and that no justification exists for the selecting of some interests (personal financial) over others (success in grants). The editorial concluded with the following statement: "The work published . . . makes no claim that the undeclared interests led to any fraud, deception or bias in presentation, and until there is evidence that there are serious risks of such malpractice, this journal will persist in its stubborn belief that research as we publish it is indeed research, not business."[10]

Five years later, *Nature,* while under the same editor, announced a change of policy in its editorial titled "Declaration of Financial Interests."[11] Its editor wrote that "there is suggestive evidence in the literature that publication practices in biomedical research have been influenced by the commercial interests of authors."[12] The new policy asked authors, before having their papers accepted, to fill out a form declaring any competing financial interests. Those who refused to disclose their interests were asked to declare that statement in writing.

JOURNAL CONFLICT OF INTEREST POLICIES

Beginning in the late 1980s, professional associations began to acknowledge the importance for scientific authors to disclose their interests in journal publications. In 1988, the International Committee of Medical Journal Editors (ICMJE) encouraged but could not require its member journals to ask authors to acknowledge any financial relationship they had that "may pose a conflict of interest."[13] Two years later, the American Federation for Clinical Research (AFCR) published its recommendation that researchers publicly disclose all research funding. Moreover, AFCR also recommended that a researcher not hold equity in any commercial entity that makes a product the researcher is investigating.[14]

The ICMJE continued to provide leadership on setting ethical standards for medical publications. In 1993, the committee acknowledged the importance of conflict-of-interest disclosure for reviewers of journal manuscripts: "The editors must be made aware of reviewers' conflicts of interest to interpret the reviews and judge for themselves whether the reviewer should be disqualified."[15] Five years later, the ICMJE affirmed the same disclosure responsibility for all medical communications: "When they submit a manuscript, whether an article or a letter, authors are responsible for recognizing and disclosing financial and other conflicts of interest that might bias their work."[16] In response to what the ICMJE sees as the growing influence of pharmaceutical companies on clinical research (particularly on the control of data and publication), the organization issued a news release in September 2001 signed by its twelve-member panel of journal editors. It stated that its affiliated journals would adopt the policy of routinely requiring authors to disclose details of their own role and of their sponsor's role in a clinical study. It also stated that some affiliated journals "will ask the responsible author to sign a statement indicating that he or she accepts full responsibility for the conduct of the trial, had access to the data, and controlled the decision to publish. . . . We will not review or publish articles based on studies that are conducted under conditions that allow the sponsor to have sole control of the data or to withhold publication."[17]

The ICMJE represents a rather small group of medical journals and is thus hardly representative of the total number of science and medical journals in publication. Moreover, the ICJME's recommendations are not routinely adopted by medical schools. In a study published in the *New England Journal of Medicine* that surveyed U.S. medical schools on how closely they complied with the ICMJE research guidelines, researchers found that "academic institutions routinely participate in clinical trials that do not adhere to ICMJE standards of accountability, access to data, and control of publication" and that "academic institutions rarely ensure that their investigators have full participation in the design of the trials, unimpeded access to trial data, and the right to publish their findings."[18]

In the late 1990s, a colleague and I began to wonder how many science and medical journals had conflict-of-interest policies for authors. And for those that did have such policies, what exactly was being disclosed? How were the editors specifically handling conflict-of-interest communications from prospective authors?

I teamed up again with L. S. Rothenberg of UCLA to investigate these questions in a study of journal publications and conflict-of-interest policies. We began the study in 1998 and chose 1997 as the base year of our analysis. With tens of thousands of scientific and medical journals published each year, we needed some way to sample the total. Although we could have taken a random sample, we instead used another criterion. We decided to get a subset of the most influential science and medical journals (as determined by the citations of articles) published in the English language that are indexed by the *Science Citation Reports*. We selected the top one thousand journals ranked according to two measurements—namely, "impact factor" and "times cited factor." In other words, we included in our study those scientific and medical journals whose articles that appeared the previous year were cited, on average, most frequently as well as the journals that had the most cumulative citations. Since many journals appeared on both rankings, we ended up with 1,396 distinct high-impact science and medical journals.

The first question we posed was: How many of these journals had conflict-of-interest policies? We learned that nearly 16 percent of these journals had such policies in 1997. We found that medical journals and the more prestigious basic science journals were more likely to have conflict-of-interest policies for authors. This conclusion was confirmed by another study in which the investigators examined the conflict-of-interest policies (in 1997) of twenty-five basic science journals and twenty-five clinical journals with the highest "immediacy index" ranking (the average number of times that an article published in a specific year within a specific journal is cited over the course of that same year).[19]

After deleting two journals that appeared on both lists, the study's investigators found that 43 percent of the forty-eight medical and scientific journals surveyed had requirements for the disclosure of an author's financial interests in published reports.[20]

Returning to our study, we found that from our original inventory of nearly 1,400 journals, over 200 had COI policies. From these, we selected a subset of all the peer-reviewed journals, which amounted to 181. We then examined the original research articles in each of these journals for author–financial interest disclosures. This task involved examining 61,134 journal articles published in 1997. Surprisingly, we found that only 0.5 percent of the more than 60,000 articles (a total of 327) had any statements listing a financial interest of authors related to the subject matter of the publication. Even more surprising to us was that nearly 66 percent of the journals (all of which had disclosure policies) had no conflict-of-interest disclosures for that year.[21]

How could that be? With so much commercialization going on in science and with what other studies have been telling us, why were there so few disclosures in the very journals where you would expect to find them? Were scientists not complying with the journals' guidelines? Did authors not understand the journals' standards for disclosure? Were the journal editors giving out different signals to authors and turning a blind eye to conflicts of interest? We followed up our study of the journals with a survey of journal editors, with the hope that this information would help us explain why there weren't more disclosures.

Perhaps the editors were given the information but chose not to publish it. When we surveyed the editors, we found that 74 percent "almost always" or "always" published author disclosure statements, thus eliminating that explanation as the reason for low rates of disclosure. Of course, we also had to consider that scientists were disclosing all that *warranted* disclosure and that the rates of disclosure were not underestimating the financial interests of authors. But that explanation flew in the face of our earlier pilot study and the research on the growth of commercialization in biomedical science. For example, in a survey of 210 respondent companies who were conducting life-science research in the United States, 90 percent had relationships with an academic institution in 1994—a significant rise from a decade before.[22] From our survey, we also had reason to believe that editors took conflicts of interests quite seriously. Thirty-eight percent responded affirmatively to the question of whether they ever rejected a manuscript based on grounds of conflict of interest (primarily or in conjunction with other factors). We concluded that poor compliance to the conflict-of-interest rules of journals was the most likely explanation for low-disclosure rates in most journals with such policies.

There is reason to believe that author compliance is not especially high with conflict-of-interest guidelines among those journals that have them. The journals neither police nor evaluate author compliance with their guidelines. By simply asking authors to disclose, many of the journal editors feel satisfied that they have fulfilled their responsibility. The rest, they think, is up to the honor system. Among the minority of science and clinical journals that have disclosure policies, considerable diversity exists in how they state the policy to prospective authors. The minimalists, such as the journal *Science*, request of contributing authors "information about the authors' professional and financial affiliations that may be perceived to have biased the presentation." Similarly, the journal *Heart* requires being informed of "any grants, business interest, or consultancy that could lead to a conflict of interest." This phrasing leaves considerable latitude for individual judgment. The author is supposed to imagine whether others might perceive an affiliation-bias in the article. This projection of how others perceive their relationship varies considerably from person to person.

Other journals apply more explicit and detailed instructions to authors. In the journal *Cancer*, the editors want all involvements disclosed regarding whether or not the authors believe others will perceive them as "biasing their outcome." They require authors to sign a disclosure affidavit that states: "By signing below, I certify that any affiliations with or involvement (either competitive or amiable) in any organization or entity with a direct financial interest in the subject matter or materials discussed in the manuscript (e.g., employment, consultancies, stock ownership, honoraria, expert testimony, and the like) are noted below. Otherwise, my signature indicates I have no such financial interest."

Finally, a third group of journals uses a template system that requires each author to check the relevant boxes in a standardized form. Therefore, all authors are on record of declaring whether they have interests or not. Among those journals that offer a template approach is the *Journal of Bone and Joint Surgery* (British and American volumes), which asks the author (after a decision has been made to accept a paper) to select one of five statements that is relevant: "Benefits are accrued from a commercial party on the subject of the article for: (1) personal or professional use only (2) personal and professional use and for a research fund, foundation or educational institution; (3) a research fund, foundation, educational institution exclusively. (4) No benefits have been or will be received. (5) Authors choose not to respond."

Not only is there an escape clause for authors who wish not to comply with disclosure, but merely selecting a statement among the five choices reveals no information about the explicit conflict. It offers readers and reviewers no help whatsoever in weighing the type of involvement for its potential bias.

A few journals—notably, the *New England Journal of Medicine*—are not satisfied that disclosure alone is sufficient to protect the integrity of the material it publishes. For a number of years, the *NEJM* prohibited the publication of certain types of articles when an author had a conflict of interest. In its instructions to authors, the *NEJM* stated: "Because the essence of reviews and editorials is selection and interpretation of the literature, the *Journal* expects that authors of such articles will not have any financial interest in a company (or its competitor) that makes a product discussed in the article." However, on June 13, 2002, *NEJM* announced that there was a change in its conflict-of-interest policy. The single word "significant" was added to its paragraph on reviews and editorials.[23] In addition, the journal announced that it will allow editorials and reviews from authors who have received honoraria over a year's period or who own less than $10,000 stock in any one company; such authors may apply the operational meaning of "no significant financial interest."

The reason given by *NEJM*'s editors for reducing the standards on conflicts of interest was that they could not find enough qualified scientists to write the articles. The editors bemoaned that their policy had reduced the population of prospective authors of editorials and reviews to nearly the null class. They wrote, "In the past two years we have been able to solicit and publish only one Drug Therapy article on a novel form of treatment."[24] They went on to say that if journals like *NEJM* do not publish reviews and editorials because of their uncompromising policy on conflicts of interest, then pharmaceutical companies will become the principal source of information about new drug therapies. In other words, their compromised policy was the lesser of two perceived evils.

Other journals had not adopted the standard that *NEJM* set; in actuality, they could only aspire to it. Plenty of opportunity still existed for academic faculty to publish journal editorials and reviews even when they had financial interests, small or large. At a time when ghost authorship of reviews and editorials in medicine is still tolerated, the *NEJM* (which has been at the vanguard of setting ethical standards in publication) believed it could not compete in the marketplace of experts when they chose the "gold standard" of ethical conduct. The deeper issue requires a change in the incentive system, one that recasts the moral infrastructure of academic medicine so that journals can have a large reservoir of independent experts who have no financial links to the products and therapies they study.

SILENT SPONSORS OF BOOKS

When the results of a major long-term study on hormone replacement therapy (which indicated that the risks outweighed the benefits) were leaked to the public days before the study's publication in *JAMA* on July 17, 2002, questions were raised about how postmenopausal estrogen therapy became so popular with American women before long-term studies had even been carried out. An estimated twenty million American women were taking doses of estrogen therapy to build stronger bones as well as reduce hot flashes and nocturnal sweats. In tracing the roots of public acceptance, *New York Times* journalists Petersen and Kolata reported that the pharmaceutical company Wyeth, a manufacturer of hormone replacement therapy drugs, financed a best-selling 1966 book *Feminine Forever*, authored by an M.D. who accented his career by promoting estrogen treatments for women. This case begs the question, How can we know that popular books with medical advice about the benefits of drugs are not the public relations projects of the manufacturers?

Historically, conflict-of-interest concerns related to books usually referred to book reviews in scientific journals, where authors of review articles rarely disclose their associations. Even the *NEJM* was caught off guard when in 1997 it permitted a critical book review on the environmental causes of cancer without citing the reviewer's affiliation as director of medicine and toxicology of a multinational chemical company.[25] (Strictly speaking, the journal's policy was not focused on book reviews, but rather on scientific review articles.)

The controversy over conflicts of interest and book reviews took another turn when an editorial was published in the journal *Addiction*.[26] The editors affirmed the journal's policy that books reviewed would include disclosures of the book author's conflicts of interest. Rarely do book authors include conflict-of-interest disclosure statements in the acknowledgment section of their work. Many scientific books are supported totally or in part by interest groups. Inadvertent disclosure may be made, however, when the author acknowledges private funding sources. Otherwise, book publishers assume no responsibility to meet an ethical standard on financial disclosure that has been adopted by many journals.

The editors of *Addiction* admitted that two books got through their screen. One book that their journal reviewed was on alcohol and health, and it received support from the liquor industry, an association that was not revealed to the readers or the reviewer. A second book, which offered a critique of the "nicotine addiction hypothesis," was reviewed in the journal without reference to the authors' involvement with the tobacco industry. A rather frank exchange took place in the pages of the journal between the editors and one of the authors. The authors were hired as experts by the tobacco industry to evaluate the thesis that nicotine was addictive. They did not see their consultantship with the tobacco industry and authorship of their book as conflictual activities: "We were paid for our time, not for our opinions."[27]

The editors of *Addiction* believed that books, no less than journals, should have ethical guidelines that make conflicting interests transparent: "Scientific books should be the undistorting mirrors of scientific truth or they have no worth."[28] The editors asked this rhetorical question to the readers of the journal: "Is *Addiction* right in believing that the ethics of academic publishing put upon authors and publishers an absolute and inalienable responsibility to declare potential or actual conflicts of interest?"[29] They argue that, as a standard practice, book publishers of scientific texts should ask for conflict-of-interest statements of authors and editors and that their books should disclose the statements. Because such practices are not in place, the editors of *Addiction* have instituted a policy that authors and editors of any book received by the journal for review will be asked to sign a conflict-of-interest statement before the review process proceeds.

The disclosure policy that has been adopted by the editors of *Addiction* should be viewed by the general public as an indicator of how intensified corporate efforts have become in influencing science. Journal editors are the gatekeepers of "certified knowledge." When they become concerned, as increasingly they have, about private interests casting shadows on academic science, then it should be a wake-up call to everyone in society to take heed. One journal editor wrote: "In biomedical publication, academic, professional, institutional, and financial interests may bias judgments and interfere with the dissemination of scientific information."[30]

Science and clinical journals as well as book publishers have considerable ground to cover to establish transparency for the conflicts of interest of authors, reviewers, and editors; and to prevent the most egregious conflicts from showing up on the pages of their journals. For example, in one case, a team of researchers investigated the reporting of conflicts of interest in guidelines for clinical practice published in journals. After sampling 191 guidelines on preventive and therapeutic interventions from six major clinical journals published between 1979 and 1999, the researchers found that only seven guidelines mentioned conflicts of interest. The authors of the study wrote: "Despite some recent improvements, reporting of conflicts of interest in clinical guidelines published in influential journals is largely neglected."[31] Even this being said, closing the disclosure loopholes will not change one important result: Academic science is losing its innocence. What once used to be the natural breeding ground of public-interest science has become the incubator of market-driven science. The next chapter discusses the loss of one of our most important social resources—namely, public-spirited science, a resource that cannot be restored by the mere transparency of interests.

NOTES

1. In 1963 estimates placed the number of scientific journals in current publication at about 30,000. See Derek J. de Solla Price, *Little Science, Big Science* (New York: Columbia University Press, 1963). Ulrich's online database in medicine and biology contained 29,683 publications in all languages in 1997. See Sheldon Krimsky and L. S. Rothenberg, "Conflict of Interest Policies in Science and Medical Journals: Editorial Practices and Author Disclosures," *Science and Engineering Ethics* 7 (2001): 205–218.

2. Sheila Slaughter, Teresa Campbell, Margaret Holleman, and Edward Morgan, "The 'Traffic' in Graduate Students: Graduate Students As Tokens of Exchange between Academic and Industry," *Science, Technology and Human Values* 27 (Spring 2002): 283–312.

3. Dorothy Nelkin, "Publication and Promotion: The Performance of Science," *The Lancet* 352 (September 12, 1998): 893.

4. Arnold Relman, "Dealing with Conflict of Interest," *New England Journal of Medicine* 310 (May 31, 1984): 1182–1183.

5. Relman, "Dealing with Conflict of Interest," 1182.

6. Marcia Angel, "Is Academic Medicine for Sale?" *New England Journal of Medicine* 342 (May 18, 2000): 1516–1518.

7. Eliot Marshall, "When Commerce and Academe Collide," *Science* 248 (April 13, 1990): 152, 154–156.

8. Sheldon Krimsky, James Ennis, and Robert Weissman, "Academic Corporate Ties in Biotechnology: A Quantitative Study," *Science, Technology & Human Values* 16 (Summer 1991): 275–287.

9. Editorial, "Avoid Financial 'Correctness,'" *Nature* 385 (February 6, 1997): 469.

10. Editorial, "Avoid Financial 'Correctness,'" 469.

11. Editorial, "Declaration of Financial Interests," *Nature* 412 (August 23, 2001): 751.

12. Editorial, "Declaration of Financial Interests," 751.

13. International Committee of Medical Journal Editors, "Uniform Requirements for Manuscripts Submitted to Biomedical Journals," *Annals of Internal Medicine* 108 (1988): 258–265.

14. American Federation for Clinical Research (AFCR), "Guidelines for Avoiding a Conflict of Interest," *Clinical Research* 38 (1990): 239–240.

15. International Committee of Medical Journal Editors, "Conflicts of Interest," *The Lancet* 341 (1993): 742–743.

16. International Committee of Medical Journal Editors, "Statement on Project-Specific Industry Support for Research," *Canadian Medical Association Journal* 158 (1998): 615–616.

17. International Committee of Medical Journal Editors, "Sponsorship, Authorship, and Accountability," *New England Journal of Medicine* 345 (September 13, 2001): 825–826.

18. Kevin A. Schulman, Damin M. Seils, Justin W. Timbie, et al., "A National Survey of Provisions in Clinical-Trial Agreements between Medical Schools and Industry Sponsors," *New England Journal of Medicine* 347 (October 24, 2002): 1335–1341.

19. The "immediacy index" is a measure of how quickly the "average article" in a journal is cited. It is calculated by dividing the number of citations of articles published in a given year by the number of articles published in that year. See www.isinet.com/isi/.

20. S. van McCrary, Cheryl B. Anderson, Jelena Jakovljevic, et al., "A National Survey of Policies on Disclosure of Conflicts of Interest in Biomedical Research," *New England Journal of Medicine* 343 (November 30, 2000): 1621–1626.

21. Sheldon Krimsky and L. S. Rothenberg, "Conflict of Interest Policies in Science and Medical Journals: Editorial Practices and Author Disclosure," *Science and Engineering Ethics* 7 (2001): 205–218.

22. D. Blumenthal, N. Causino, E. Campbell, K. S. Louis, "Relationships between Academic Institutions and Industry in the Life Sciences—An Industry Survey," *New England Journal of Medicine* 334 (February 8, 1996): 368–373.

23. Jeffrey M. Drazen and Gregory D. Curfman, "Financial Associations of Authors," *New England Journal of Medicine* 346 (June 13, 2002): 1901–1902.

24. Drazen and Curfman, "Financial Associations of Authors," 1901.

25. Jerry H. Berke, review of *Living Downstream: An Ecologist Looks at Cancer and the Environment* by Sandra Steingraber, *New England Journal of Medicine* 337 (November 20, 1997): 1562.

26. G. Edwards, T. R. Babor, W. Hall, and R. West, "Another Mirror Shattered? Tobacco Industry Involvement Suspected in a Book Which Claims That Nicotine Is Not Addictive," *Addiction* 97 (January 2002): 1–5.

27. Edwards et al., "Another Mirror Shattered?" 2.

28. Edwards et al., "Another Mirror Shattered?" 1.

29. Edwards et al., "Another Mirror Shattered?" 4.

30. Annette Flanagin, "Conflict of Interest," in *Ethical Issues in Biomedical Publication,* ed. Anne Hudson Jones and Faith McLellan (Baltimore: The Johns Hopkins University Press, 2000), 137.

31. George N. Papanikolaou, Maria S. Baltogianni, Despina G. Contopoulos-Ioannidis et al., "Reporting of Conflicts of Interest in Guidelines of Prevention and Therapeutic Interventions," *BMC Medical Research Methodology* 1 (2001): 1–6. E-Journal: biomedcentral.com/1471-2288/1/3.

11

THE DEMISE OF PUBLIC-INTEREST SCIENCE

It is a common mistake to speak of "the mission" of a modern research university. Universities have several missions, just like they have diverse disciplines and distinctive schools. The missions of the modern university are derived in large part from the university's multiple personalities. For this chapter, I have identified four personalities emblematic of the research university, each of which has a distinctive relationship with knowledge and a form or structure uniquely designed to realize its missions.

The first personality, the *classical form* of the university, is best characterized by the Aristotelian expression "Knowedge is virtue." The primary emphasis of the classical tradition is teaching, basic research, and critical examinations of cultural traditions of civilization. Knowledge is pursued for its intrinsic value. Science is nourished by the free and open exchange of information. Research paths are investigator-driven, rather than imposed by funding sponsors. Proprietary or classified research is prohibited. The university in its classical form is not beholden to anyone, and it is not designed to serve any social or political purpose. Science serves no master and is bound only by the norms of universal cooperation and the methods of rational inquiry.

The second personality is associated with the writings of the seventeenth-century British philosopher-scientist Francis Bacon. The *Baconian ideal* of the university is best understood through the expression "Knowledge is productivity," a derivative of Bacon's aphorism "Knowledge is power." Under this ideal, the primary focus of the university is to provide the personnel, the knowledge, and the technology for economic and industrial development. The pursuit of knowledge is not fully realized until it has contributed to the industrial economy. Terms like "technology transfer," "intellectual property," and "university-industry partnerships" are part of the lexicon that expresses the new relationships between the fruits of academic science and industrial growth. The responsibility of the scientist begins with discovery and ends with commercial application. Universities exist mainly to provide labor for industry and to help industry turn knowledge into technology; technology into productivity; and productivity into profits.

The *defense model* of the university derives largely from the role universities have played during wartime, where the phrase "Knowledge is security" defines academia's relationship to the country's national defense effort. Government-sponsored research laboratories and the scientists who manage them are viewed as a national defense resource. Congress created the Defense Advanced Research Projects Agency (DARPA) under the Department of Defense to support academic research directed at improving military capabilities. During World War II, radar, sonar, and the atomic bomb were high priorities for the defense program; during the Vietnam War, universities provided research on the control of popular insurgency against brutal dictators in developing nations, on biocides in jungle environments, and on stealth weapons. More recently, the themes for domestic-security research include missile-defense systems, protection from biological and chemical weapons, and combating terrorism. Universities that restrict classified research, prohibit contracts for weapons development, reject the Reserve Officer Training Corps (ROTC), or refuse CIA funding rebuff the defense model, preferring to accentuate their other three personalities. With passage of the Homeland Security Act in 2002, new sources of funding for university research on domestic security and terrorism will provide incentives for the defense model to reemerge in academia after years of withdrawal during the post–Vietnam War period.

Finally, I call the fourth personality the *public-interest model*, which is best understood by the phrase "Knowledge is human welfare." In this model, one of the university's primary functions is to solve major societal problems in health and human welfare, such as dread disease, environmental pollution, and poverty. Professors are viewed by others as well as themselves as a public resource called upon to tackle complex medical, social, economic, and technological problems. A university that embraces public-interest science may view its role as *agent provocateur*, nurturing its faculty to advance public welfare by investigating inequities, researching the cause of disease, and forecasting the impacts of new technologies. Programs like the "war against cancer" were based on the premise that research funds should be targeted to universities under a focused mission to best serve the public interest (i.e., to cure a disease; or in this case, a class of diseases). University researchers are recruited into the service of public-interest science by the vast outlays of the federal government and, to a lesser degree, private foundations.

My argument in this chapter is that the relative decline in publicly supported science, in conjunction with the commercialization of universities, has brought about a decline in public-interest science. By this statement, I mean that entrepreneurial scientists no longer identify themselves as having a commitment to investigate public-interest problems per se. The choice of problems is dictated by commercial and not social priorities. In other words, the Baconian ideal of the university has become the dominant personality.

Biologist and Nobel laureate Philip Sharpe commented that "as universities become more identified with commercial wealth, they also lose their uniqueness in society. They are no longer viewed as ivory towers of intellectual pursuits and truthful thoughts, but rather as enterprises driven by arrogant individuals out to capture as much money and influence as possible."[1] In their book *To Profit or Not to Profit,* which examines the transformation of nonprofit institutions toward a for-profit orientation, Powell and Owen-Smith state, "The changes underway at universities are the result of multiple forces: a transformation in the nature of knowledge and a redefinition of the mission of universities by both policymakers and key constituents. These trends are so potent that there is little chance for reversing them—nor necessarily a rationale for doing so."[2] Speaking at an Emory University conference on the commercialization of the university, Harvard's President Emeritus Derek Bok commented that "by legitimating profit-seeking, commercialization [of universities] puts the voluntary communal spirit, that set of values that makes professors spend more time preparing their courses, working with students, going to faculty meetings, in danger; professors will spend more time consulting and starting companies."[3]

The principal discussions around the trends of academic entrepreneurship have focused on issues of trade secrecy, intellectual property, and conflict of interest. Among these issues, the losses to the scientific profession are at the centerpiece of debate over policy responses. Also, there is a recurrent discussion of the erosion of public trust in multivested science. One of the most serious losses, however, is ignored, and that is the loss of public-interest scientists.

ACADEMIC CAPITALISM

The evolving academic universe is no longer as nurturing an environment for public-interest science as it once was. To a large degree, universities have been taken over by money managers and academic entrepreneurs who are looking for financially lucrative research. This scenario usually translates to research that will result in intellectual property. Conversely, research that reveals degradation of our natural resources, that exposes charlatan claims of companies, or that investigates the environmental causes of disease usually offers no financial benefits to the university. Many of these studies are done with little or no overhead value, which universities have increasingly come to expect from research faculty.

In 1998, a group of scientists from diverse disciplines gathered in Woods Hole, Massachusetts, and began a new professional group called the Association for Science in the Public Interest (ASIPI). Among the stated principles of ASIPI is that scientists "whose training and professions are subsidized by the public have an ethical and professional obligation to use their skills and expertise in the service to the public."[4] "Public-interest science" is therefore defined as "research carried out primarily to advance the public good."

Professors at universities who spend their time incubating venture capital spin-offs from research will likely have neither the time nor the inclination to devote themselves to public-interest work. They must be involved in writing patent applications, preparing business plans, communicating with potential investors, working with a management team, learning about stock offerings, and meeting the demands of the regulatory matrix in which their private enterprise functions. Many universities began to convince themselves that unless they became enterprise zones, they would lose their edge. Princeton University, arguably one of the last strongholds of the classical model, has recently approved a policy that allows the university to take minority ownership in the start-ups of its professors.[5]

For their book *Academic Capitalism,* Slaughter and Leslie interviewed scientist–entrepreneurs from four countries. They confirmed their suspicion that academic scientists will change in value structure when they embrace market activities. They wrote: "We would expect that faculty as professionals participating in academic capitalism would begin to move away from values such as altruism and public service, toward market values."[6] The academic faculty they interviewed did just that. But they also did something else. They redefined the role of "public-interest science" so that it was consistent and not discordant with entrepreneurial activities. "Rather they elided altruism and profit, viewing profit making as a means to serve their unit, do science, and serve the common good."[7]

The mind-set of entrepreneurship is in distinct opposition to the mind-set of public purpose. The former embraces the values of social Darwinism with its attendant features of competition and survival. One company must win the patent before another company gets there first. Protection of trade secrecy is essential to succeed in a competitive climate. Regulations are viewed as an impediment to success. They slow down progress and create inefficiencies in getting products to market.

The latter mind-set is based more on a holistic sensibility. We are all in this together. There are communitarian values. The university has a special role to play in examining the critical needs of society, apart from the market system, particularly for the most vulnerable groups who do not have a voice. How can science help these groups?

That question is not the one asked by someone starting a university-based company. Public-interest science addresses issues that elude a market solution. One cannot invest in the study of occupational illness and expect to make a profit. The closest thing we have are tort lawyers. Lawyers will take cases, such as occupational illness, on contingency, where they hope to make a profit from a positive jury verdict. But the law firm does not fund scientific research. The research into occupational illness is largely publicly funded. The mission behind corporate funding of occupational disease is to defend the safety of a product. Occupational health science is primarily public-interest oriented because the questions investigated are connected to preventing disease; the questions are not directed at a treatment, where the commercial interests lie.

Public-interest science asks how knowledge can contribute to ameliorating social, technological, or environmental problems. Private-interest science asks how knowledge can produce a profitable product or defend a corporate client, whether or not it has social benefits and whether or not the product is distributed fairly and equitably. Even in cases where taxpayers have funded years of research, scientists can turn a discovery built on such research into a profitable drug. Many people, whose very taxes funded the research and who are in dire need of the drug, now cannot afford it.[8]

The irony cannot be avoided. Capitalism is supposed to operate on the principle that private risk yields private loss or private wealth. Philanthropy turns private wealth into social resources. But the idea that public risk (that is, publicly supported research) should be turned into private wealth is a perversion of the capitalist ethic. Under this scheme, government funds research; and when a discovery is made, it is turned into private wealth for the scientist, the university, and the company partner. The university becomes a key player in turning public funds into private wealth. The scenario is justified by the claim that the discovery and its benefit to society would not have occurred if these incentives were not in place. Whether this rationale is valid is a matter of debate. Not all discovery is turned into a public good under these conditions.

There has been a major shift in university management. Central administration has invested heavily in technology transfer and in developing the apparatus to sell the development rights of faculty discoveries. Moreover, the ideology of technology transfer at universities has been used to redefine public-interest science. Rhodes and Slaughter wrote: “Generally, administrators articulated an organizational ideology that accounted for university involvement in technology transfer in terms of the public interest as manifested in regional (private) economic development."[9] It is argued that when academic scientists act as entrepreneurs, maximizing their personal wealth, they act in the public interest. The public interest is met, according to this view, precisely when market forces are introduced into universities. According to this position, as long as university patents rise and products are produced, the public interest is served. One economic indicator of how the new technology-transfer policies are working is the number of new patents taken out by universities. No one examines, however, what kinds of products are produced, who gets served, and how the profits are distributed. Moreover, who is investigating how commercialization is changing the academic research agenda? I have previously emphasized that investment into studying some of society's most serious problems is not lucrative to the university and that it may even be costly to the industrial system in the short-term while benefiting society in the long-term. It is these problems and the commitment to them by academic researchers that are lost in the new business climate of academia.

I have argued that the new wave of entrepreneurial science has produced a series of what has been generally understood as "manageable problems." Donald Kennedy, in a *Science* editorial, spoke of "patent disputes, hostile encounters between public and private ventures, and faculty distress over corporate deals with their universities."[10] In addressing conflicts of interest, what universities, journals, and government agencies do is impose disclosure requirements. And in response to the problem of secrecy, universities establish time limits whereby scientific results can be sequestered before they are allowed to be published (e.g., ninety days or six months). However, the loss of a public-interest ethos within science is incalculable and not amenable to remediation by degree.

Many academic scientists who apply their intellect to the analysis and solutions of public problems make contributions that neither government nor industry scientists are in a position to make. The following are profiles of the lives and contributions of three individuals who have exhibited a commitment to public-interest science. While their career paths may be discordant with the current trends of the new entrepreneurial university, their work underscores the value of free and independent inquiry, which is not beholden to and skewed by the patronage of private interests.

BARRY COMMONER

Barry Commoner is widely regarded as one of the most influential environmental scientists of the twentieth century. He devoted much of his sixty-year scientific career to investigating issues of paramount social concern, including radioactive fallout, nuclear power safety, the causes of pollution, the impact of a petroleum-based economy, the dangers of dioxin from waste incinerators, and the relationship between ecology and economics.

The son of immigrant parents, Commoner was born in 1917 in Brooklyn, New York. While a preadolescent, his interest in science was piqued when his uncle, a Russian-born intellectual on the staff of the New York Public Library, gave him a small microscope and some educational guidance. Commoner was a high academic achiever at James Madison High School, where he spent considerable time in biology labs. His teachers advised him to go to college to study biology, which was not an assumed path for the children of immigrants at that time. Because his uncle understood that Jews had a difficult time getting jobs in academia, he advised his nephew that if he had any intentions to pursue a career in higher education to apply to an elite college—specifically not City College, the default choice of many immigrant children. Commoner applied to Columbia University and was initially denied admission despite his stellar academic record. The admissions office directed him to apply to Seth Low Junior College, which Columbia had established to accept ethnic minorities (Jews, Italians, and Blacks) who they declined for admission.

Commoner rejected the default college. His aunt, a well-known poet, protested his rejection to Columbia through a prominent professor at the university who argued the case before the administration. Commoner's rejection was overturned. He entered Columbia in 1933 during the height of the Great Depression. By the time he was enrolled, Commoner became personally aware of the conditions of inequity in higher education, which he understood to be a reflection of broader societal injustices. He recalled that he "entered college with a very sharp and clear picture of a society in trouble."[11] While he pursued his studies in science at Columbia, Commoner became active on current issues of the 1930s, including the Scottsboro case (young black men in the South convicted of murder on false charges) and poverty in Harlem. At the time, he viewed science as being a separate and pure subculture operating within a flawed society: "I was brought up to believe that science was a thing with its own independent history, with its own ideals, its own purposes, its own goals, as an autonomous objective entity."[12]

From Columbia, Commoner pursued graduate studies in biology at Harvard. He became active in the Association of Scientific Workers, a group that took on issues of international politics. He left Harvard in 1940 with his degree requirements completed, and he accepted a position at Queens College, where he stayed for a year and a half. At the start of the Second World War, Commoner volunteered for the Navy. He was assigned to the Navy's experimental air squadron group at Patuxent Naval Air Station, outside of Washington, D.C. One of his tasks was to devise a delivery system for planes so that they could spray DDT over enemy territory in the Pacific to protect America's invading ground troops from tick-borne diseases.

His entrée into politics occurred when Commoner was assigned by the Navy to temporary duty with the Senate Military Affairs Subcommittee, headed by Harley M. Kilgore. The Navy was interested in pending legislation to establish the National Science Foundation. During this period, Commoner began to witness the interplay between science and politics in the debates about control over the National Science Foundation and the future of atomic energy. He helped organize the first congressional hearings at which atomic scientists spoke out about the perils of atomic weapons.

After the war, in 1947, Commoner accepted a position at Washington University in St. Louis, where his early research focused on spectroscopic analysis of single cells. Through the influence of Warren Weaver of the Rockefeller Foundation, Commoner became active in the American Association for the Advancement of Science (AAAS) for the purpose of getting the organization interested in relating science to social problems. He served as chair of the AAAS Committee on Science and the Promotion of Human Welfare. Commoner sought ways to implement his idea that scientists had a responsibility to inform the public on critical issues of public policy. Collaborating with leading scientists like Linus Pauling and Edward Condon, Commoner participated in a petition campaign to mobilize scientists against atmospheric testing and to alert the public of the dangers of radioactive fallout. The role of scientists, he believed, was to bring information to citizens, who could then reach their own conclusions. The role of scientists was not to act as elite decision makers with a monopoly over technical knowledge.

Commoner believed that to be effective as a public-interest scientist, first and foremost you had to demonstrate that you can do good science. His own contributions to basic science in the fields of genetics, plant virology, and biochemistry can be found in leading academic journals. He established two large research programs: one on free radicals, which is credited with the first discovery of free radicals in living systems,[13] and a second on the biochemistry of tobacco-mosaic virus replication.

Commoner continued to implement his ideas about translating and communicating science to the citizenry by helping to organize the St. Louis Committee for Nuclear Information (CNI) in 1958 and by contributing to a newsletter called *Nuclear Information*, which evolved into the magazine *Scientist and Citizen* (later changed to *Environment* in 1969). The CNI sought ways to educate the public about the biological dangers of radioactive fallout from atmospheric testing of nuclear weapons during the 1950s.

In one of the campaigns organized by the CNI, parents of young children were asked to deliver their children's baby teeth to the organization. Once collected, CNI sent the teeth for laboratory tests to determine their level of strontium 90, one of the common radioactive by-products of the fallout from atmospheric testing. The School of Dentistry at Washington University collaborated with CNI and established an analytical laboratory to measure the strontium 90 absorption in the children's teeth. Describing the campaign, Commoner wrote, "Citizens have contributed significantly to what scientists now know about fallout. Through the St. Louis Baby Tooth Survey, the children of that city have contributed, as of now, some 150,000 teeth to the cause of scientific knowledge about fallout."[14] By 1966, the Baby Tooth Survey collected over 200,000 teeth.[15] The survey and the worldwide petition of scientists helped to solidify President Kennedy's support behind the 1963 Nuclear Test Ban Treaty. Commoner noted, "The Nuclear Test Ban Treaty victory was an early indication of the collaborative strength of science and social action. It was this conclusion that led CNI to become the Committee for Environmental Information and extend its mission to the environmental crisis as a whole."[16]

A second example, discussed in Commoner's book *The Closing Circle,*[17] illustrates the role of creative scientific thinking to resolve contested issues in environmental policy. In the mid-1960s, the Director of Public Health in Decatur, Illinois, contacted Commoner about the high-nitrate concentrations found in the city's water supply. Decatur, a city of about 100,000 people and located 120 miles from St. Louis, drew its drinking water from the Sangamon River, which winds through a region rich in corn and soybean farms. The director informed Commoner that the nitrate levels of the river exceeded federal standards every spring. He wanted to know whether the high-nitrate levels were coming from fertilizers (which were used heavily in this plush agricultural area) or were coming from the soil, the preferred explanation by the agricultural community. He challenged Commoner to use his scientific knowledge to solve a public health problem, whereupon the Center for the Biology of Natural Systems, which he had founded in 1966, took up the study of Decatur's nitrate problem.

Commoner asked himself: How can the source of nitrogen contamination be found? Recalling his experimental work on the isotopes of nitrogen, Commoner reasoned that there would be different ratios of isotopes ^{14}N to ^{15}N in soil and in nitrogen fertilizer. He knew that the nitrogen in ammonia fertilizers have an abundance of the ^{15}N isotope compared to the soil, where the nitrogen comes from the breakdown of humus and other organic materials.

Using a mass spectrometer to determine the nitrogen–isotope ratios, Commoner's team of scientists collected water that was draining from the fertilized fields into the Sangamon River, and they measured the relative abundance of the two isotopes in the samples, which they then compared to the ratios they found in the soil and in synthetic fertilizers. The analysis showed that approximately half of the nitrogen in the city's drinking water was traced to artificial nitrogen fertilizers. Their results stimulated a lively and sustained discussion about the use of nitrogen fertilizer in the surrounding farms, eventually leading to the regulation of agricultural fertilizers in the surrounding farms.[18]

A third example of public-interest science carried out in Commoner's center, after it had moved from Washington University in St. Louis to Queens College in New York City, was a study done in the mid-1990s on the dispersal of dioxins. The work was driven by a series of questions pertaining to dioxin contamination of isolated nonindustrial regions of North America. Why is the Inuit population that lives in the Arctic Circle getting so much exposure to dioxin? Why do women in the Nunavut Territory of the Arctic Circle have twice the level of dioxin in their breast milk as women in the United States or lower Canada? From what sources is the dioxin coming?

Commoner and his colleagues collected emissions data on 44,091 dioxin sources in North America during a period between 1996 and 1997. They obtained weather pattern data, and they applied a computer model, adapted from earlier uses for tracking radioactive leaks in nuclear plants, to predict the movement of airborne dioxin emitted from each of these sources. Commoner's group found that essentially all of the dioxin deposited on ecologically vulnerable sites in Nunavut came from sources outside the province, most of them from the United States (70 to 80 percent). Relatively few sources are responsible for most of the dioxin deposited on Nunavut; only 1 to 2 percent of the 44,091 sources account for 75 percent of the deposition.[19] As one of the nation's leading scientists on dioxin contamination, Commoner led a grass roots campaign in New York City against the administration's plan to build eight incinerators for burning municipal garbage.

• • • • •

We often hear about the competitive nature of science, particularly in pursuit of discovery and more recently in protecting commercially valuable intellectual property. Public-interest science is directed at problems and hypotheses for which there is very little competition because there is no pot of gold at the end of the investigation. Quite the contrary, public-interest scientists may face SLAPP (Strategic Lawsuit Against Public Participation) suits, criticism from peers for bringing science into the public arena, and disparagement from government entities who act defensively in the face of evidence that discredits their ability to protect public health.

The next scientist profiled devoted his career to studying public health threats from environmental contaminants.

HERBERT NEEDLEMAN

In his thirty-five-year career in academic medicine and public health, Herbert Needleman has been a major protagonist and scientist pioneer behind one of the most successful campaigns in disease prevention of the past century—the removal of lead from the environment. Needleman's career path took him to three universities, where he devoted his medical research largely toward understanding the health effects of lead on children and where he focused his public-interest science on removing the sources of lead contamination.

While preparing this profile of him, I attended a talk he delivered at a conference on lead poisoning held at a public housing development in South Baltimore, sponsored by the Baltimore City Department of Social Services. Speaking to a largely African-American audience of child and family service providers, Needleman explained that the hazards of lead contamination had been suspected for over 2,000 years but that the first systematic studies were reported only sixty years ago. He told the audience that African-American children are at especially at high risk of lead poisoning, an environmental contamination that has severely damaging effects on learning abilities and behavior.

Herbert Needleman was born to middle-class parents in Philadelphia, Pennsylvania, just a few years before the Great Depression. His father earned his livelihood as a furniture merchant, pursuing business opportunities in several cities. Needleman's interest in becoming a doctor began when he was ten years old. By the time he was ready for college, his parents had moved to Allentown, Pennsylvania, for business purposes. Needleman attended Muhlenberg College as a commuter student and majored in general science. After Muhlenberg, he attended the University of Pennsylvania Medical School, where he chose pediatrics as his medical specialty, mostly because he admired the physicians he had met while on his rotating internship.

To help pay his way through medical school, Neddleman took a summer job as a laborer at a DuPont chemical plant in Deepwater, New Jersey, the state where tetraethyl lead was synthesized years before. Workers were not allowed to smoke or carry matches in the plant. Twice a day, a smoking whistle blew (at 10:00 A.M. and 2:00 P.M.), when hundreds of workers left the plant to go to wooden smoking shacks in open areas. He observed that many workers seemed clumsy, remote, and out of touch. A smaller number of workers spent their break time listlessly staring into space. He learned from other people at the plant that many workers suffered the effects of lead poisoning for years.

Needleman was drafted into the Army between 1955 and 1957. He was stationed at Fort Meade, Maryland, where he became chief of pediatrics and supervised the treatment of "preemies," an experience that reinforced his desire to become a pediatrician. Upon discharge, he completed his pediatric residency, and after a short stint working with middle-class families, he became more interested in at-risk populations. He describes two pivotal career-shaping experiences he had as a young M.D.

In 1957 while a resident in pediatrics, a seasoned colleague of his examined a three-year-old Hispanic girl and diagnosed her as having acute lead poisoning. Needleman learned that the girl got her lead poisoning from chips and dust from lead-based household paint. Then he did what he was taught to do in such a circumstance—namely, apply chelation therapy. The young girl, who had been close to comatose, began to get better when the lead was removed from her blood. Needleman told the girl's mother that she had to move; otherwise, her daughter would be re-exposed and again become seriously ill. The woman told him that wherever she moves, it will be the same. There was no way for her to escape the lead. At that point, he realized that it was not enough to make a diagnosis and prescribe medication. The disease of lead poisoning was in the living experience of the afflicted child. Something had to be done to change the environment.

In a second case, he recalled a talk he gave to adolescent youth in a Black church in Philadelphia. When a young boy approached him after his talk and began discussing his ambitions, it took but a brief conversation before Needleman realized that the boy was brain damaged. This incident reminded him of the time as a resident he treated the lead-poisoned child; it got him to think more seriously about how little has been done to prevent lead toxicity. He remembered seeing many young Black children peering out the windows of their tenements while he took the trolley to the hospital. He wondered how many of them were not attending school because of lead poisoning. Ironically, it was because of low admission rates of young poor children that the Department of Mental Health began to do lead screening in schools.

When the National Institute of Mental Health provided fellowships for young M.D.'s to enter psychiatry, Needleman decided to take a residency in that field, thinking that he might enter child psychiatry. From that point on, his work was devoted to the effects of environmental toxins on children's mental and physical well-being. His introduction to public-interest medicine started in the 1960s when he provided consultation to a community legal services lawyer who had teamed up with an organization of welfare-rights mothers to work with a New York State congressman for the passage of the first federal law on lead paint.

Needleman began to study lead contamination of children in 1970 when he was an assistant professor of psychiatry at the Temple University Medical Center in Philadelphia. He was influenced by a 1943 paper by a Harvard pediatric neurologist, Randolph Byers, which suggested that some cognitive dysfunction of school children may be the result of lead toxicity.[20] Byers discovered that children who were referred to him for evaluation of aggressive behavior were formerly patients with lead-poisoning symptomology. Later, Needleman learned that Byers was threatened with a million-dollar lawsuit by the Lead Industries Association for suggesting that environmental exposures to lead could be endangering children.

In searching for a good stable biological marker of lead contamination, Needleman eliminated blood and urine. He considered hair as a potential marker, which was easy to collect. But he realized hair was subject to environmental contamination and therefore not a trustworthy indicator of lead inhalation. Prompted by a reference in the literature, he came upon the idea of using baby teeth for assessing lead exposure. He obtained the teeth of suburban and inner-city children in Philadelphia through Temple and University of Pennsylvania dental clinics and through a suburban pediodontist; he then tested a subsample of that group. In a follow-up study, Needleman collected teeth from children enrolled in the Philadelphia school system in an area known by locals as the "lead belt" because of its proximity to the old lead industry. Needleman found that those with elevated lead in their teeth scored lower on intelligence tests, speech, and language function. His results were published in 1972 in a 700-word article in the journal *Nature*[21] and in 1974 in the *New England Journal of Medicine*.[22]

The publications in *Nature* and the *New England Journal of Medicine* were the starting points of Needleman's foray into public policy and public-interest science. On the basis of that 1972 paper, he was invited by the EPA to present a talk at an international scientific meeting held in Amsterdam. That experience awakened him to the intense politics of lead. Needleman wrote of the event:

> I was unprepared by my past attendance at pediatric meetings for what I encountered there. This was not scholarly debate on the toxicology and epidemiology of lead; this was war. . . . Arrayed against each other were a small and defensive group of environmentalists and health scientists on one side, and on the other the representatives of the gasoline companies, including such formidable

> entities as EI DuPont, Associated Octel, Dutch Shell, and Ethyl Corporation of America. Any paper suggesting that lead was toxic at lower doses immediately faced a vocal and well-prepared troop that rose in concert to attack the speaker.[23]

His paper was not spared.

Needleman realized that much of the research on lead toxicity was paid for and controlled by the industry "who had a tight grip on what the public was permitted to know."[24] He decided he had to study the issues on its own terms without influence from the industry that profits from lead.

Continuing his work at the Harvard Medical School from 1971 to 1981 and collaborating with the top-ranking epidemiologist Alan Leviton, Needleman refined his studies with more sophisticated statistics and better controls. He analyzed the teeth of twenty-five hundred first and second graders from two similar Boston suburbs. He compiled information on the academic status of the students and investigated whether there was an association between high lead levels and poor academic performance. His results, published in the *New England Journal of Medicine* in 1979, showed a direct link between low chronic levels of lead exposure and impaired mental development. The paper generated a flurry of media attention. The lead industry requested and was denied his raw data. By that point, Needleman had become the prime target of the lead industry's counterattack.

Meanwhile, in the late 1970s, the EPA was preparing an air standard for lead. Needleman became involved in reviewing the air–lead criteria document, initially through the request of the Natural Resources Defense Council and subsequently as a contractor for the EPA. His findings concluded that elevated dentine lead levels in children were associated with decrements in IQ. When this information began to influence public policy, industry started to attack the integrity of his studies. These attacks intensified during 1990 when an attorney from the Department of Justice asked Needleman to participate in a landmark suit brought under the Superfund Act against three lead polluters in Utah.

Two scientists, hired by the lead industry, requested and were given approval by Needleman to see his raw data. They spent two days in his lab taking notes from his lab notebooks. The federal suit was settled before it went to trial. The government was awarded $63 million for the cleanup of the mining site. Based on the notes of their paid scientist–consultants, the lead industry made a formal complaint to the NIH's Office of Scientific Integrity accusing Needleman of scientific misconduct. The NIH asked the University of Pittsburgh, where Needleman has been appointed professor of psychiatry and pediatrics since 1981, to undertake an investigation of his research. During the period of the investigation, which began in October 1991, his research files were locked by order of the University of Pittsburgh. He was allowed to view his data only in the presence of the scientific integrity officer of his university. The university panel investigating the lead industry charges found no evidence of fraud, falsification, or plagiarism. But curiously, when the panel issued its findings in December 1991, it left a residue of suspicion by stating that the panel members were not able to exclude the possibility of "misconduct in terms of misrepresentation." They were responding to his accusers, who argued that the models he used in his analysis were selected to maximize a lead effect. The panel members, however, offered no evidence for their suspicion. Unsettled by the panel's enigmatic findings, Needleman pressured the university for an open hearing of his case in a venue where world-class experts were allowed to participate. When the faculty assembly passed a unanimous resolution and Needleman filed a complaint for such a hearing in federal court, the university capitulated and the open hearing was held. Independent scientists who reanalyzed the 1979 data reached virtually the same conclusion. In May 1992, a hearing board at the University of Pittsburgh, by unanimous decision, found no evidence of scientific misconduct.

In honor of his work, Needleman received the second Heinz Foundation Environmental Award for his "extraordinary contributions to understanding and preventing of childhood lead poisoning." The award statement of the Heinz Foundation noted that he "worked tirelessly and at great personal cost to force governments and industry to confront the implications of his findings. While this has made him the target of frequent attacks, he has fought his critics with courage, tenacity and dignity."

Much of Herbert Needleman's academic career in medical and public health research has been devoted to studying the effects of lead on children and to communicating those effects to policy makers. As part of his public-interest science, Needleman recently coauthored a book for concerned parents titled *Raising Healthy Children in a Toxic World.* He has also served as adviser to a number of public-interest groups, including Citizens for a Better Environment, Center for Science and the Public Interest, Mothers and Others for Pesticide Limits, and Mount Sinai's Center for Children's Health and the Environment.

The university setting, Needleman confided, was the only place where the type of public health research to which he has been committed could be accomplished in the face of industry threats and political pressures. The universities where Needleman had his appointments were sufficiently insulated from external influences and were supportive of his academic freedom so as not to interfere with his scientific and public-interest work.

In the third profile of this series, a young scientist combines her passion for neuroscience with her commitment to improve the environmental health of disenfranchised urban communities.

LUZ CLAUDIO

If you ask most scientists about their public-interest work, they are likely to respond by reciting a list of either advisory committees on which they have served or talks they have given in nonprofit venues. As commendable as these contributions may be, there is another meaning of public-interest science that has its roots in the citizen-science initiatives of the 1960s, the science advocacy organizations of the 1970s, and the environmental justice movement of the 1990s. The concept of public-interest science that emerges from these traditions is "science by and for the people," where "the people" is often a euphemism for at-risk or underrepresented communities. To fully realize this idea, scientists must form partnerships with communities, just as entrepreneurial academic science forms partnerships with the private sector. Taking the analogy one step further, as with corporate-sponsored research, residents in community-centered research become involved in framing and implementing a study of local problems.

Luz Claudio comes as close to fulfilling this public-interest role as anyone I have encountered. In a discussion we had at her office at the Department of Community and Preventive Medicine at Mount Sinai School of Medicine, I learned much about Claudio's scientific career, which brought her from a small mountain village in Puerto Rico to one of the nation's preeminent research centers in environmental and occupational medicine. The medical center is situated on 101st Street in Manhattan, just on the border between the Upper East Side, one of the wealthiest neighborhoods in the world, and Spanish Harlem, one of the poorest neighborhoods in New York.

Claudio traced her early interest in nature to the influence of her grandmother, who she describes as an indigenous "Curandera," the Spanish word for curer. Her grandmother was an enthusiastic collector of wild plants, which she classified and studied for their medicinal properties. As a young child, Claudio was her student and her offtime "volunteer" for medicinal remedies and natural beauty treatments. Her grandmother provided the inspiration for Claudio's early interest in observing nature, at a time when science, for Claudio, was still a distant and unarticulated goal.

As a student in middle school, Claudio was a high achiever who relished the intellectual challenge of science courses ("The periodic table was the most amazing thing I had ever seen") and who demonstrated a natural ability in mathematics. Although she completed middle school with high grades, Claudio's teachers steered her to a vocational high school, where she was advised to acquire secretarial skills so that she could find a job as soon as she graduated. But after doing poorly in typing, she persuaded her parents to have her transferred to the other high school in her region, which offered a general education and therefore provided a path to college.

While in high school, Claudio worked part-time at a urologist's office, where she eventually took on responsibilities for lab technician's work, which whetted her appetite for pursuing studies in medicine and biology. After high school, she attended one of the satellite campuses of the University of Puerto Rico (at Cayey), which was conveniently located near a small rain forest. Graduating in 1984, Claudio majored in biology with an emphasis on ecology and natural systems. After completing a postcollege internship in Hawaii, where she worked at a prawn aquacultural farm, Claudio applied to and was accepted into a doctoral program at the Albert Einstein College of Medicine. She fulfilled her doctoral work in 1990, specializing in neuropathology by investigating how immune cells cross the blood-brain barrier, a central process in multiple sclerosis.

After receiving her Ph.D., Claudio was awarded a fellowship by the American Association for the Advancement of Science, where she completed a paper (subsequently published[25]) on in vitro methods that could be used to test the neurotoxicity of chemicals—an area of research prompted by regulatory concerns over the effects of chemicals on the developing fetus. With the experience of this fellowship behind her and with her vivid recollections of contaminated communities in Puerto Rico, Claudio made a decision to focus her career on environmental issues. In 1991, she was hired by Phil Landrigan, who heads the Division of Environmental and Occupational Medicine at Mount Sinai School of Medicine, to link environmental medicine with neurobiology. The program at Mount Sinai was founded by the late Irving Selikoff, the legendary occupational physician who uncovered asbestos disease among miners and shipyard workers.

After four years at Mount Sinai, Claudio was made director of the Community Outreach and Education Program, a move that reflected her division's commitment to addressing environmental health concerns of minority communities. Incentives for such programs came in the aftermath of President Bill Clinton's 1993 Environmental Justice Executive Order, which required federal agencies to take account of disparate risks to communities in their regulatory decisions; it also came in response to the new community orientation of the National Institute of Environmental Health Sciences under Kenneth Olden's leadership. This position gave Claudio the opportunity to develop new models of public-interest science that were neither elitist nor paternalistic.

When Claudio received a grant from the Environmental Protection Agency to teach community leaders about environmental medicine, she brought activists together from Brooklyn, the Bronx, and East Harlem and asked them to talk about the environmental health concerns in their neighborhoods. Expecting them to bring up lead paint, she instead heard, to her surprise, a collective response that focused on a growing incidence of asthma in their communities. The community leaders worked with Claudio to develop a methodology to investigate asthma cases in the city and to ascertain whether the rates differed across neighborhoods. She used a data set on citywide hospitalizations to ask the question: What were the community rates of asthma-related hospitalizations in the city of New York? The values of asthma hospitalization rates were mapped by zip code, divided into shaded quintiles according to levels, and correlated with household income. The study found that the highest values for hospital admission rates were in the Bronx, also the poorest borough in the city. They also found that the zip code areas with the largest minority populations had the highest asthma hospitalization rates. The lead sentence in the coverage of the study in the *New York Times* read: "In the first study of its kind of the nation's urban asthma epidemic, researchers have found the rate of hospitalization from the disease far greater among children in urban poor, predominantly minority neighborhoods of New York City than experts suspected."[26]

The study provided community health advocates the scientific grounds for petitioning government to strengthen air quality standards, to improve preventative and therapeutic medical care for underserved populations, and to prohibit new sources of air pollution in communities that already bear the burden of high asthma rates.

Claudio's asthma prevalence study was the stimulus for an environmental health-impact analysis that followed a similar model of community-based science. In the late 1990s, Consolidated (Con) Edison of New York sold for development the highly valued property on which its midtown electric generating plant stood. To compensate for the loss of electricity from the dismantling of this plant, Con Edison planned to add more generating power to its lower East Side plant on Fourteenth Street. The lower East Side plant would thus burn more fuel and emit more emissions. A community organization, called the East River Environmental Coalition, contested the permit for the new plant operations on environmental justice grounds. They argued that this increase would impose more risk to a community already at risk from air pollution. Initially, there was suspicion, but no science, behind the concerns.

Working with lower East Side community leaders, Claudio designed a study to determine the air pollution sensitivity of the residents who lived in close proximity to the plant. The research team conducted a survey of over four hundred individuals who resided in four buildings near the plant, and they found an asthma prevalence rate of about 23 percent (with active asthma at 13 percent)—more than double the national rate. They also found chronic bronchitis rates in this population at two-and-a-half times the national average. This study produced evidence to community leaders that some residents would be disproportionately affected by higher emission rates from the scaled-up energy generation plant. Claudio's model of community-based science also included coauthorship of publications by Mount Sinai scientists and community leaders involved with the study. The findings of the survey were used by lower East Side community organizers to negotiate the abatement of toxic emissions, air monitoring, and community involvement in an ongoing advisory committee to Con Edison.

Community-based public-interest science, like that conceptualized and implemented by Claudio and other scientists, has helped residents in New York City neighborhoods respond to health threats from waste to energy plants, diesel engine exhaust, incinerators, highway expansion, and many other urban developments that create greater risks to low-income minority communities.

• • • • •

Without committed public-interest academic scientists who are willing and capable of involving community leaders in research, all the technical and scientific expertise would be monopolized by one group of stakeholders, thus leaving the residents with strongly held but largely unexamined intuitions about their health risks. Universities that have internalized the ethos of academic capitalism will close off opportunities for the invaluable contributions to public life illustrated by the three public-interest scientists profiled in this chapter on the grounds that more economic value to the institution can be derived from space allocated to commercial partnerships.

NOTES

1. Philip A. Sharpe, "The Biomedical Sciences in Context," in *The Fragile Contract: University Science and the Federal Government,* ed. D. H. Guston and K. Keniston (Cambridge, Mass.: MIT Press, 1994), 148.

2. Walter W. Powell and Jason Owen-Smith, "Universities As Creators and Retailers of Intellectual Property: Life Sciences Research and Commercial Development," in *To Profit or Not to Profit,* ed. Burton A. Weibrod (Cambridge, U.K.: Cambridge University Press, 1991), 192.

3. Derek Bok, "Barbarians at the Gates: Who Are They, How Can We Protect Ourselves, and Why Does It Matter?" A talk given at Emory University, at the Sam Nunn Bank of America Policy Forum titled "Commercialization of the Academy," April 6, 2002.

4. Website of the Association for Science in the Public Interest: public-science.org (accessed August 31, 2002).

5. Jospeh B. Perone, "Phi Beta Capitalism," *The Star-Ledger,* February 25, 2001.

6. Sheila Slaughter and Larry L. Leslie, *Academic Capitalism* (Baltimore: The Johns Hopkins University Press, 1997), 179.

7. Slaughter and Leslie, *Academic Capitalism,* 179.

8. Jeff Gerth and Sheryl Gay Stolberg, "Drug Companies Profit from Research Supported by Taxpayers," *New York Times,* April 23, 2000.

9. Gary Rhodes and Sheila Slaughter, "The Public Interest and Professional Labor: Research Universities," *Culture and Ideology in Higher Education: Advancing a Critical Agenda,* ed. William G. Tierney (New York: Praeger, 1991).

10. Donald Kennedy, "Enclosing the Research Commons," *Science* 294 (December 14, 2001): 2249.

11. Barry Commoner, interview by D. Scott Peterson, April 24, 1973.

12. Barry Commoner, interview by D. Scott Peterson, April 24, 1973.

13. B. Commoner, J. Townsend, and G. E. Pake, "Free Radicals in Biological Materials," *Nature* 174 (October 9, 1954): 689–691.

14. Barry Commoner, "Fallout and Water Pollution—Parallel Cases," *Scientist and Citizen* 7 (December 1964): 2–7.

15. Barry Commoner, *Science and Survival* (New York: Viking Press, 1971), 120.

16. Barry Commoner, an interview with Alan Hall, *Scientific American* 276 (June 23, 1997): 1–5, at sciam.com/search/search_result.cfm (accessed November 29, 2002).

17. Barry Commoner, *The Closing Circle* (New York: Alfred A. Knopf, 1971).

18. D. H. Kohl, G. B. Shearer, and B. Commoner, "Fertilizer Nitrogen: Contribution to Nitrate in Surface Water in a Corn Belt Watershed," *Science* 174 (December 24, 1971): 1331–1333.

19. Barry Commoner, P. Bartlett et al., "Long Range Transport of Dioxin from North American Sources to Ecologically Vulnerable Receptors in Nunavut, Arctic Canada," final report to Commission for Environmental Cooperation (October 2,000).

20. R. K. Byers amd E. E. Lord, "Late Effects of Lead Poisoning on Mental Development," *American Journal of Diseases of Children* 66 (1943): 471.

21. H. L. Needleman, O. Yuncay, and I. M. Shapiro, "Lead Levels in Deciduous Teeth of Urban and Suburban American Children," *Nature* 235 (1972): 111–112.

22. H. L. Needleman, E. M. Sewell, and I. Davidson et al., "Lead Exposure in Philadelphia School Children: Identification by Dentine Lead Analysis," *New England Journal of Medicine* 290 (1974): 245–248.

23. Herbert L. Needleman, "Salem Comes to the National Institutes of Health: Notes from Inside the Crucible of Scientific Integrity," *Pediatrics* 90 (December 6, 1992): 977–981.

24. Needleman, "Salem Comes to the National Institutes of Health," 977.

25. Luz Claudio, "An Analysis of U.S. Environmental Protection Agency Testing Guidelines," *Regulatory Toxicology and Pharmacology* 16 (1992): 202–212.

26. Holcomb B. Noble, "Far More Poor Children Are Being Hospitalized for Asthma, Study Shows," *New York Times,* July 27, 1999, B1.

12

PROSPECTS FOR A NEW MORAL SENSIBILITY IN ACADEMIA

Many people, I among them, believe that American universities have gone too far: they permit themselves and their faculty to become ridden with conflicts of interest; and through their aggressive support of technology transfer and their liberal acceptance of industry contracts, they compromise the integrity of scientific inquiry and communication. While subfields in medicine and the biosciences have been at the epicenter of these debates, other disciplinary voices can also be heard. In assessing the commercial trends in their field, two geographers wrote: "we should also not forget how universities have simultaneously served as places from which to question the status quo and imagine better futures. It is this role as relatively autonomous spaces of critical reflection and resistance that today seems increasingly eclipsed by the marketization of education, the commodification of knowledge, and the simple but relentless pressures to produce."[1]

Some presidents, like University of Michigan's Duderstadt, believe that faculty and administrators are, by and large, honorable and responsible people "who will behave properly in balancing the university's interests and their own responsibilities for teaching and research against their interests in intellectual property development and technology transfer."[2] Like other apologists for the aggressive commercialization of academia, Duderstadt believes that "the key to avoiding conflict of interest is public disclosure."[3] In actuality, disclosure simply provides a rationalization for continuing to create more serious conflicts, that is, as long as universities are open about them.

Consider the case of John Mendelsohn, a distinguished professor of oncology at the M. D. Anderson Cancer Center at the University of Texas. In the early 1980s, Mendelsohn discovered an antibody (C225) that seemed to inhibit tumor growth by blocking a molecular signal pathway. The company ImClone developed an anticancer drug called "Erbitux" from the antibody. As prospects for Erbitux grew, ImClone signed a $2 billion contract with Bristol-Myers Squibb to market the drug.

Mendelsohn joined ImClone's scientific advisory board and had accumulated stock in the company at one time worth $30 million. The M. D. Anderson Center was one of the sites for the clinical trials of Erbitux.

The results of the clinical trials did not get the approval for Erbitux ImClone sought from the FDA, and controversy ensued. ImClone was accused of multimillion-dollar stock trades by company insiders weeks before FDA's negative decision was disclosed. It was also alleged that the FDA granted fast-track status to Erbitux based on the wrong version of a research protocol. And the founder and CEO of ImClone faced criminal charges for insider trading and issuing false statements.

During his appearance before a House subcommittee on Energy and Commerce, Mendelsohn was queried about his conflict of interest with ImClone and whether he violated any ethical standards. Mendelsohn replied: "I have never treated a patient with C225 [Erbitux]. . . . Whenever I have given scientific talks or written papers or have public meetings, I have always stated my holdings in the company and my memberships on its scientific advisory committee and on its board."[4] Mendelsohn also noted that his affiliation with the company appears on consent forms at M. D. Anderson. Is this form of disclosure adequate to address the conflicts of interest described by Mendelsohn? Was he at sufficient arms length from any influence he might have had on the testing at Erbitux? Given his involvement with ImClone, should the drug have been tested at his institution?

Signals abound that the pendulum has swung too far, that disclosure will not be accepted as the universal antidote for the conflict of interest, and that some restraints are being considered. The moral boundaries of research and publication integrity have been tested, and efforts are underway to reset the moral barometer to a higher level. There are five major stakeholders that are beginning to react to the public's negative attitudes about conflicts of interest in science and medicine. They are the journals, the professional societies, the governmental agencies, the universities, and the nonprofit research institutes. I begin by looking at the trends among journals in addressing authors' conflicts of interest.

SCIENTIFIC PUBLICATIONS

Over the past decade, the scientific journals have become more sensitized to the appearance of conflict of interest by their contributors and among editorial staff.

During the 1990s, many journals adopted conflict-of-interest policies. Most policies added financial disclosure requirements for authors. But some journals took a tougher stance and sought to prevent authors from contributing editorials or reviews if they had a direct conflict of interest with the subject matter. The medical journals are far ahead of the pure science journals in establishing author guidelines on conflict of interest. The issue of ghostwriting, prestige authorship, and authors who do not have full control of the data they are publishing is troubling to some journals. The International Committee of Medical Journal Editors took a bold step in recommending that their member journals get a signed statement that the author alone makes the decision to publish the data and that the individuals listed as authors have made a worthy contribution to the article.

Compliance with the COI policies of academic journals is based on the honor system. Most journals do not have the personnel or the time to assess the level of compliance. When my colleague and I published a study of the rates of author disclosure in 181 peer-reviewed journals with COI policies, a number of editors were surprised that their journals did not have a single disclosure of personal financial interest for the entire test year. In the current climate of academic commerce, it seemed unlikely that authors had nothing to disclose. This improbable result raises the question of compliance when a policy has no sanctions against violators.

Except for a small number of internationally acclaimed journals that do well financially, many operate under tight fiscal margins. Electronic access to journals has reduced the demand for hard-copy subscriptions, placing some journals in even greater economic peril. If editors place too much emphasis on author conflicts of interest, they may lose potential contributors to a journal that doesn't require the disclosure of their personal income.

While the journals are making some effort to respond to the rising tide of scientific conflict of interest, they cannot do it alone. Except for a few journals that proscribe certain authors from publishing editorials or review, most can only offer voluntary disclosure. The response to authors' conflict of interest must be systemic, and it must include all players in the research community.

PROFESSIONAL SOCIETIES

Several professional associations in science and medicine have begun to develop policies on conflicts of interest. The American Medical Association (AMA) has been a leader in addressing the problems of COI since 1990. As the largest medical association for American physicians, the AMA has published several principles and supporting opinions that establish its code of ethics. The AMA's *Principles of Medical Ethics*, revised in 1980, represents its primary guidelines from which all other codes and recommendations are derived. Among its seven principles, however, is no reference to conflict of interest. In 1990, the AMA adopted six principles under the heading "Fundamental Elements of the Patient–Physician Relationship." One of these principles states: "Patients are also entitled . . . to be advised of potential conflicts of interest that their physicians might have and to secure independent professional opinions."[5] The AMA's *Code of Medical Ethics* (*CME*) is a compilation of opinions based on the interpretation of its adopted principles and from reports applied to specific cases. In the section devoted to biomedical research, it states: "Avoidance of real or perceived conflicts of interest in clinical research is imperative if the medical community is to ensure objectivity and maintain individual and institutional integrity."[6] It also states that medical researchers who receive funding from a company "cannot ethically buy or sell the company's stock until the involvement ends" and that clinical investigators are obligated to disclose their financial conflicts of interests in all published results, including letters.[7] Because these statements are solely guidelines, we have no basis for knowing what percentage of the AMA membership complies with its principles.

Other professional societies have issued policy statements on conflicts of interest, particularly those groups that are involved with human-subjects research. The American Society of Gene Therapy, which adopted a statement titled "Financial Conflicts of Interest in Clinical Research" on April 5, 2000, set its standards higher than the federal guidelines: "All investigators and team members directly responsible for patient selection, the informed consent process and/or clinical management in a trial must not have equity, stock options or comparable arrangements in companies sponsoring the trial."[8]

Federal guidelines issued in 1995 set the standard for disclosure for researchers at equity ownership in companies exceeding 5 percent and/or aggregate payments received from companies in excess of $10,000 per year. The guidelines do not prohibit any activity or research relationship, but they do leave it to the individual institutions to manage a "significant COI."

The Association of American Medical Colleges (AAMC) is the professional group representing 125 accredited U.S. medical schools, covering also 400 teaching hospitals and approximately 90,000 medical school faculties. The AAMC's views on conflicts of interest were issued in its 1990 publication, *Guidelines for Dealing with Faculty Conflicts of Commitment and Conflicts of Interest in Research.* Like many of its brethren university associations, the AAMC supports the basic principles behind the partnership between industry and academia as "essential to preserve medical progress and to continue to improve the health of our citizenry."[9] But the AAMC also asserts that "the mere appearance of a conflict between financial interests and professional responsibilities may weaken public confidence in the researcher's objectivity." How do they reconcile these views that some relationships are essential but that they have an *appearance* that weakens the public's trust?

Maintaining that human-subjects research requires an especially high standard of moral integrity, the AAMC applies the regulatory concept of "rebuttable presumption" to establish the burden of proof: "Institutional policies should establish the rebuttable presumption that an individual who holds a significant financial interest in research involving human subjects may not conduct such research."[10] Moreover, the AAMC maintains that the principle—namely, "the rebuttable presumption against significant financial interest in human subjects research"—should apply whether the funding is public or private.

Under the AAMC's ethical code on conflict of interest, the institution may provide compelling evidence—presumably to itself, since this is a voluntary code—that even with significant financial interests, the investigator may be permitted to conduct the research.[11] This position harkens back to the federal ethics guidelines for advisory committees, where the first principle is to prevent scientists with COIs to participate and the second principle is to permit waivers for the scientists with COIs when there are "compelling circumstances." The approach taken by the AAMC is to establish a high standard but to leave plenty of latitude for exceptions. Like the federal agencies, the AAMC states: "When the individual holding such interests is uniquely qualified by virtue of expertise and experience and the research could not otherwise be conducted as safely or effectively without that individual, he or she should be permitted the opportunity to rebut the presumption against financial interests by demonstrating these facts to the satisfaction of an institution's conflict of interest (COI) committee."[12]

Just how effective will institutional conflict-of-interest committees be in the face of large grants and contracts, and researchers with weighty résumés? Sample cases exist that raise questions about the power of institutional oversight groups against imperial faculty (see the Fred Hutchinson Cancer Research Center case later in this chapter; pp. 209–210). But progress has been made, as demonstrated by the AAMC, which recommends that institutions report COI information to the university's Institutional Review Board (IRB), a federally mandated body. The IRB is responsible for approving human-subject protocols, and some IRBs are already taking on the additional responsibility of making judgments about conflicts of interest.

We continue to hear the argument that the management of conflict of interest and human-subject protection should be kept distinct. However, after the death of Jesse Gelsinger (see chapter 8), there has been more discussion about folding COI into informed consent. Lawsuits filed on behalf of victims who died in clinical trials have argued that failure to disclose a COI or to prevent a COI introduces additional risks to the patient, which should either be prevented or disclosed.

The Association of American Universities established a special task force on research accountability to produce a report and issue recommendations on individual and institutional financial conflict of interest. The task force was cochaired by Steven B. Sample and L. Dennis Smith, the presidents of the University of Southern California and the University of Nebraska at Lincoln, respectively. Also included in the task force were the presidents of the University of Iowa, Princeton University, and Columbia University. The AAU published the task force's findings in October 2001.[13] After reviewing the available information, the task force concluded that "although the definitive data about the prevalence of conflict of interest is lacking, academic-industry relationships are clearly increasing, and with them, the risk of conflicts of interest compromising the integrity of research conducted in academia continues to rise."[14] While generally favoring an institutional, case-by-case approach to conflicts of interest among university faculty, the task force issued a special warning for situations involving human subjects and presented a zero-tolerance recommendation: "Since research involving humans creates risks that non-human research does not, any related financial interest in research involving humans should generally not be allowable."[15]

But like the AAMC, the task force tempered its zero-tolerance recommendation by leaving open the opportunity for exceptions. "If compelling circumstances justify an exception to this general rule, the research should be subject to more stringent management measures."[16] Both the AAU and the AAMC positions ask the universities to avoid, wherever possible, financial interests in human trials, to be able to defend it if necessary, and to apply the ethical rules regardless of the source of funds. In addition, both believe that the IRBs should be involved in reviewing or monitoring conflicts of interests. The AAU asserts rather confidently that the institutional IRB has jurisdiction over whether a particular financial interest should be managed and/or disclosed to the human subjects. This jurisdiction is not, however, in the legislation and charter of the IRBs. Many IRBs are not equipped to handle these types of questions. The AAU task force recommended a double layer of protection against financial conflicts in human-subjects experiments. The first layer of review should be made by the institutional conflict-of-interest committee. The recommendation of this committee should then be passed on to the IRB, which also makes its determination. According to the AAU task force, "In such a system, neither the

IRB nor the conflict of interest committee would be able to override the other's management requirements if the result would be to lessen the stringency of the management requirements. Either one could prohibit the research from proceeding, unless the financial conflict of was removed or mitigated."[17]

The AAU task force report provides the strongest safeguards to date proposed by any review committee on managing conflicts of interest involving human subjects. The recommendations are designed, however, to keep the government's role to a bare minimum, to avoid further federal regulations, to foster the principle of self-governance, to keep flexibility within the institutions, and to establish ethical principles that extend beyond the bare minimum of federal regulations. For example, the government has not issued any restrictions on institutional financial conflict of interest, while the AAU task force reported that this type of COI "strikes to the heart of the integrity of the institution and the public's confidence in that integrity."[18] The major categories of institutional financial conflicts of interest are twofold: one involves university equity holdings or royalty arrangements and their affect on both research programs and those programs involving university officials who have personal financial interests in faculty companies; the second involves trustees whose firms could supply the university with goods or services. What is at stake?

In the task force's words, "Institutional conflicts can reduce a university's role as an objective arbiter of knowledge on behalf of the public."[19] The AAU issued warnings and general guidelines to the institution, but no clear prohibitions. For institutional conflicts of interest, the AAU task force cited three principles: always disclose; manage the conflicts in most cases; and prohibit the activity when necessary to protect the public interest or the interest of the university. But how reasonable is it to expect that universities will choose the higher moral ground of public interest when the institution's bottom line or acquiescence to its big donors is at stake? In the past ten years, several of the large professional societies have begun to address the serious erosion of academic integrity for both the individuals' and the institutions' conflicts of interest, which have become endemic to the university culture. The specialized medical groups have not responded as well. A study of twenty-one medical association guidelines published in 2000 by the Office of the Inspector General of DHHS found that only two associations (the American College of Emergency Medicine and the American Psychiatric Association) had codes of ethics with explicit reference to the disclosure by physicians to their patients/subjects of financial conflicts of interest.[20] The federal government's response has also been slow and cautious, as they have made some incremental changes in response to the demands for more accountability.

FEDERAL AGENCY CONFLICT OF INTEREST POLICY CHANGES

In the wake of high-profile scandals and internal investigations, which began in the late 1990s, federal agencies have begun to tighten up on their conflict-of-interest policies. The trend has been toward greater transparency, more comprehensive disclosures of financial relationships to funding and regulatory agencies, and the protection of the self-management of COIs by grantee institutions. The interest of government is twofold. It wants to project an image of fairness and objectivity in policy decisions; it also wants to assure the general public that money going into research, especially clinical research, is not being tainted by the real or perceived bias of scientists who have a financial interest in the subject matter of their research.

The U.S. General Accounting Office, an investigative arm of government, reviewed the selection of external advisers on EPA panels in its June 2001 report. The GAO found that conflicts of interest were not identified and mitigated in a timely manner and that the public was not adequately informed about the points of view represented on its influential panels of experts. By June of 2002, the EPA drafted its new guidelines with changes in how it would deal with conflict of interest.

The EPA's Scientific Advisory Board (SAB) consists of an administrative staff and about one hundred experts in various fields of science. When advisory panels are chosen to review an area of science that informs a policy, the panel is constituted by members of the SAB, with additional experts chosen at large. According to provisions of the Ethics in Government Act of 1978, the SAB staff screens panel candidates for conflicts of interests and the appearance of lack of impartiality.

The new policy has several changes that will be more attentive to conflicts of interest and that will afford the public more involvement in the process through which individuals are recommended to serve on expert panels. According to the new policies and procedures, "If a conflict exists between a panel candidate's private financial interests and activities and public responsibilities as a panel member, or even if there is the appearance of partiality, as defined by federal ethics regulations, the SAB Staff will, as a rule, seek to obtain the needed expertise from another individual."[21] Previously, a prospective panel member filled out a COI form annually. Under the current standards, they file a COI form each time they are recommended to serve on a SAB panel. The forms that must be filed by panel members are also more detailed. Each prospective panel member is asked to write in narrative form any relationship they have that could be perceived as a conflict of interest. If a COI is identified, the staff must consult with the chair of the SAB Executive Committee. Moreover, the public is given the opportunity to protest potential candidates for a panel on grounds of bias. The EPA has the legal authority to waive the conflict of interest of a potential panel member; however, by making the process more transparent to the public, the EPA will have to think more seriously about exercising the waiver since it requires documentation.

The Public Health Service and the National Science Foundation issued COI regulations in 1995 for all recipients of grants from those agencies. This step was a significant one toward establishing a reporting mechanism and a responsible party for managing COIs within each institution that receives federal grants.

As of 1997, under its *Guidelines for the Conduct of Research in the Intramural Research Program*, the NIH had a full disclosure policy for its scientists that includes a statement on all relevant financial interests, including those of the scientist's immediate family. The disclosure is required to any funding agencies before the scientists' participation in peer review, to meeting organizers before any presentation of results, and finally in all publications and communications, written and oral. The requirement that intramural scientists have to disclose all personal financial interests to journal editors does not translate, however, to extramural research (grants outside the institutes).[22] Government grants do not impose a requirement that grantees publish their results in journals that have conflict-of-interest policies. Such a requirement would undoubtedly create an incentive for more journals to adopt such policies.

Because of the public sensitivity to conflicts of interest in clinical trials and the national priority to protect the integrity of experiments with human subjects, the FDA became a lightning rod for criticism of clinical investigators who had commercial agendas. The agency published a rule in February 1998 (which became effective the following year) requiring disclosure of financial interests of clinical investigators that could affect the reliability (efficacy or safety) of data submitted to the FDA by sponsors of applications for product approval. The FDA rule requires that anyone who submits a marketing application of any drug, biological product, or device must submit information about the financial interests of any clinical investigator who made a significant contribution to the demonstration of safety. Failure to file the information could result in FDA's refusal to approve the application, or it may require additional tests.

The FDA requires certification from the applicant that no financial arrangements with a clinical investigator have been made where the outcome of a study could affect the compensation. In other words, companies should not be giving investigators payments in stock options or special rewards for drugs that "work." Investigators should have no proprietary interest in the tested product, and the investigator should not have a "significant equity interest" in the sponsor of the study. Thus, the FDA has adopted a few prohibitions of the most egregious conflicts of interests. For others, it requires financial disclosure, presumably so the agency can factor in conflict of interest in assessing the reliability of the data.

In May 2000, DHHS Secretary Donna E. Shalala outlined her intention to issue additional guidance to clarify its regulations on conflict of interest. Among her goals was to get the NIH and the FDA to work together "to develop new policies for the broader biomedical research community, which will require, for example, that any researchers' financial interest in a clinical trail be disclosed to potential participants."[23] One of Shalala's first initiatives was to get IRBs to address conflicts of interests for investigators and institutions. Shalala wrote that the DHHS will "undertake an extensive public consultation to identify new or improved means to manage financial conflicts of interest that could threaten the safety of research subjects or the objectivity of the research itself."[24]

It was Shalala's intention to seek new legislation to enable the FDA to issue civil penalties against violations of COI disclosure. Shalala hoped to close the loop between COI, human-subject protection, and IRBs. The failure to disclose a COI was viewed as either an increased risk or an increased perceived risk to a potential human volunteer. The DHHS issued a "Draft Interim Guidance" document in January 2001.[25] (Strictly speaking, a guidance document is not a regulation, but commentators viewed it as another layer of federal controls on research centers and universities.) It said explicitly that "HHS is offering this guidance to assist Institutions, Clinical Investigators, and IRBs in their deliberations concerning potential and real conflicts of interest, and to facilitate disclosure, where appropriate, in consent forms."[26] The draft interim guidance document proposed that the IRBs should be involved in identifying and managing both individual and institutional conflicts of interest and that consent forms for clinical trials should contain the sources of funding of the clinical investigators. A number of commentators were critical of the ill-defined concept of "institutional conflict of interest" and questioned whether IRBs could make decisions based on this concept. Would human volunteers act differently if they learned that a clinical researcher or their institution had a financial stake in the outcome of the experiment? Would it be more difficult to recruit volunteers for such studies? Could the IRBs, already overworked in many cases, be able to handle this added responsibility? These were some of the concerns raised by university trade associations such as the AAMC and the AAU to the DHHS draft interim guidance document. A second draft of the guidance document, published in the *Federal Register* on March 31, 2003, retained many of the basic recommendations of the earlier guidance document.[27] It suggested that institutions engaged in federally-supported human subjects research separate responsibilities for financial decisions and research decisions, establish conflict of interest committees (COICs), extend the responsibility of the COIC to address institutional financial interests, and use independent organizations to hold or administer the institution's financial interest.

Notwithstanding the initiatives within the Public Health Service and the FDA to manage COIs, the OHRP stated that "there is currently no uniform comprehensive approach to consideration of potential financial conflicts of interest in human research."[28]

As of early 2003, the U.S. government has taken no new formal initiatives to address institutional and researcher COI, notwithstanding the fact that universities are increasingly investing in companies founded by one or more of their faculty. In a survey of member institutions taken by the Association of University Technology Managers, 68 percent reported holding equity in companies that fund research in those institutions.[29] Faculty members at five California universities alone founded three hundred biotechnology companies.[30] The responses taken by the FDA and the NIH have largely amounted to public relations efforts responding to a climate that demands political correctness. But the response does not get to the core of the problem, which is the increasing commercial face of American universities.

ACADEMIC INSTITUTIONS

American universities are still on the learning curve with respect to faculty conflicts of interest. And with respect to institutional conflicts of interest, each university is going at it alone as a result of no federal mandate and no set of national guidelines. The most significant changes have arisen only in response to federal mandates that require research universities who are receiving NSF or NIH funding to establish a conflict-of-interest management plan. Beyond that mandate, some universities are introducing the topic of conflicts of interest in training programs on scientific integrity designed for doctoral students and clinical investigators.

The research on university COI policies thus far shows a lack of specificity and a wide variation in the content and the management of the policies. For example, one study surveyed one hundred institutions with the most funding from the NIH in 1998.[31] The policies were examined from August 1998 to February 2000. The study found that 55 percent of the policies required disclosure from all faculty, while 45 percent required them only from the principal investigators. A relatively small number of the policies (19 percent) set explicit limits on faculty financial interests in corporate-sponsored research; a mere 12 percent contained language on what type of delay in publication was permissible;

and 4 percent prohibited student involvement in work sponsored by a company for which the faculty had a personal financial interest. The study cited the need for uniform guidelines across academia. "Wide variation in management of conflicts of interest among institutions may cause unnecessary confusion among potential industrial partners or competition among universities for corporate sponsorship that could erode academic standards."[32]

When universities manage conflict-of-interest cases, they rarely ask researchers to forego those interests. Investigative journalist David Wickert reviewed thousands of financial disclosure forms at the University of Washington. He found that the university required researchers to give up all financial interests a mere eight times in 321 cases that went through the university's review process.[33]

Recently, significant attention at university medical schools has focused on clinical trials and conflict of interest. The DHHS held hearings on the subject in the aftermath of the death of Jesse Gelsinger, who was a subject in a human gene therapy experiment that failed (see chapter 8). One study (published in 2000) analyzed policies governing conflicts of interest at ten U.S. medical schools that received the largest amount of research funding from NIH. One promising sign was that five of the schools had disclosure policies that exceeded the federal guidelines requiring disclosure of financial interests. In addition, six of the medical schools required disclosure to the IRB as well as to the assigned administrator on COI policies. Four had stricter requirements than the federal government for researchers who were conducting clinical trials.[34]

Another survey, which yielded 250 responses from medical schools and research institutions, found that 9 percent had policies that exceeded federal guidelines.[35] This finding is an indication that universities' moral compasses are pointing in the direction of favoring more stringent COI policies (more than those *de minimus* standards of the federal government) to protect the values and integrity of the university.

With regard to industry contracts, universities are learning from the experiences of UCSF and the University of Toronto that restrictive covenants give the private sector control over the data or publication. Universities are increasingly refusing to accept contracts with clauses that restrict the right of academic researchers to exercise control over the research method or the publication of the results.[36] Yale's Giamatti wrote, "As an indispensable condition to arrangements for cooperative research with industry . . . the university [Yale] will not accept restrictions, inhibition, or infringement upon a member of the faculty's free inquiry or capacity orally to communicate the results of his or her research. . . . The university will not accept any restriction of written publication, save the most minor delay to enable a sponsor to apply for a patent or a license."[37]

When the faculty member is both the investigator and the corporate head, then the decision to delay publication is not an external control on the university but a part of the academic norm to maximize economic value before communicating the results of scientific research. The norm of trade secrecy arises as much from within the university as from outside. Delaying publication or denying data to other researchers can have adverse social consequences if it delays the potential therapeutic uses, if it restricts the development, or if it prevents dissemination of knowledge that might illuminate premarketing or prelicensing risks of a technique or product.

A 2001 commentary in the *Journal of the American Medical Association* discussed a few of the more promising proposals—which go beyond disclosure—for addressing both institutional and faculty conflicts of interest in academia. Two M.D.'s—Hamilton Moses III of the Boston Consulting Group and Joseph Martin, Dean of the Harvard Medical School—suggested creating an entity separate from the university that would handle all equity held by the institution and by its faculty in companies that fund basic or applied research at that institution. This plan is analogous to government appointees' putting their equity holdings into a blind trust. The authors of the proposal feel justifiably uneasy about the current manner in which universities manage their conflicts of interest. They wrote:

> Ownership of even a small amount of stock in a small publicly traded or private company, whose outcome can depend on the outcome of a clinical trial or access to a key laboratory discovery, has proven difficult to manage. It is arguable that no amount of oversight can fully prevent the inevitable appearance or reality of conflict, which may operate at both conscious and unconscious levels to sway an individual's judgment, even among those with the best of intentions.[38]

The Moses–Martin proposal, if implemented effectively, could eliminate even the appearance that private or institutional investment decisions could be linked to scientific judgments. In addition, creating a "firewall" between research and equity holdings would dispel any allegations of "insider trading." The proposal does not address other types of conflicts of interest, however, such as consulting and gifts to researchers.

INDEPENDENT RESEARCH INSTITUTES

Many of the leading research institutes receive the majority of their funding from government sources. They also typically have ties to universities where researchers have academic appointments. One of these institutions, which has been the target of ongoing investigations by government agencies and investigative journalists, is the Seattle-based Fred Hutchinson Cancer Research Center, a tax-supported nonprofit with ties to the University of Washington. Scientists at "The Hutch," as it is known by locals, were financially involved with companies that financed their research. Since it was founded in the 1970s, an estimated twenty scientists working at "The Hutch" left to start companies with equity value of more than $18 billion.[39] A series of stories in the *Seattle Times* under the headline "Uninformed Consent" brought these conflicts of interest to the attention of the general public. For many years, physicians at "The Hutch" were not required to tell patients when they had private financial interests in drugs or medical products. Clinical investigators could supervise clinical trials while they had substantial equity in a company that had a financial interest in the trial's outcome. The *Seattle Times* focused its report on two experiments, one involving a bone marrow transplant protocol, which was carried out between 1981 and 1993, and a second series of experimental treatments for breast cancer, which took place between 1991 and 1998. According to the report, an unusually high number of deaths accompanied these experimental treatments.

A physician at "The Hutch," who served on the institution's Institutional Review Board (IRB), contacted federal authorities about what he alleged were egregious conflicts of interest at his institution. The whistleblower was quoted in the *Seattle Times*: "In essence, financial conflict of interest led to highly unethical human experimentation, which resulted in at least two dozen patient deaths. Oversight committees were misled, lied to and kept uninformed while in an atmosphere of fear and intimidation."[40] According to records received by the investigative journalists, in addition to the clinical investigators, "The Hutch" itself had a financial stake in the experiments.

Ironically, the board of trustees of "The Hutch" adopted a conflict-of-interest policy in 1983 that banned employees from participating in research in which they or their family members had an economic interest of any type. However, the policy was not enforced in the cases investigated. Some scientists even claimed they did not know about the policy.

As an outgrowth of the federal investigations and media attention, "The Hutch" introduced a new conflict-of-interest policy in May 2002 that markedly restricts a clinical researcher from participating in human-subjects research if it is sponsored by a for-profit entity or designed to test a product or service of a for-profit entity in which the researcher (or his/her family) has a prohibited financial interest, which includes ownership interest of any amount or any nature in that for-profit entity. This new, rather complex policy issued by "The Hutch" signals a reform in conflict-of-interest policies taken by many of the leading research institutes and medical centers. However, the COI disclosures at these institutes and centers are internal to the institutions and are thus generally shielded from the Freedom of Information Act. Without public disclosure, failure to implement COI procedures cannot easily be detected unless there is litigation. The next chapter reaches beyond disclosure in seeking solutions to the management and prevention of conflicts of interest in academic and publicly supported research institutes.

NOTES

1. Noal Castree and Matthew Aprake, "Professional Geography and the Corporatization of the University: Experiences, Evaluations, and Engagements," *Antipode* 32 (2000): 222–229.
2. James J. Duderstadt, *A University for the 21st Century* (Ann Arbor: University of Michigan Press, 2000), p. 144.
3. Duderstadt, *A University for the 21st Century*, p. 144.
4. U.S. House of Representatives, Committee on Energy and Commerce, Subcommittee on Oversight and Investigations, *House Hearing on the ImClone Controversy*, October 10, 2002.
5. Council on Ethical and Judicial Affairs, American Medical Association, *Code of Medical Ethics* (Chicago: AMA, 2000–2001), xiii. Based on reports adopted through June 1999.
6. American Medical Association, *Code of Medical Ethics,* 69.
7. American Medical Association, *Code of Medical Ethics,* 69–70.
8. American Society of Gene Therapy, "Policy/Position Statement: Financial Conflict of Interest in Clinical Research," April 5, 2000, at www.asgt.org/policy/index.html.
9. The Association of American Medical Colleges, Task Force on Financial Conflicts of Interest in Clinical Research, *Protecting Subjects, Preserving Trust, Promoting Progress* (December 2001), 3.
10. Association of American Medical Colleges, *Protecting Subjects,* 7.
11. Association of American Medical Colleges, *Protecting Subjects,* 7.
12. Association of American Medical Colleges, *Protecting Subjects,* 7.
13. Task Force on Research Accountability, Association of American Universities, *Report on Individual and Institutional Financial Conflict of Interest* (October 2001).
14. Task Force on Research Accountability, *Report on Individual,* 2.
15. Task Force on Research Accountability, *Report on Individual,* 4.
16. Task Force on Research Accountability, *Report on Individual,* 4.
17. Task Force on Research Accountability, *Report on Individual,* 6.
18. Task Force on Research Accountability, *Report on Individual,* 10.
19. Task Force on Research Accountability, *Report on Individual,* 12.
20. June Gibbs Brown, Office of the Inspector General, Department of Health and Human Services, *Recruiting Human Subjects* (June 2000), OEI-01-97-00196, appendix A, pp. 16–17.

21. Policies and Procedures Subcommittee, Executive Committee, Science Advisory Board, U.S. Environmental Protection Agency, *Overview of the Panel Formation Process at the EPA Science Advisory Board* (May 2002), A4.

22. National Institutes of Health, *Guidelines for the Conduct of Research in the Intramural Research Program at NIH,* adopted January 1997, at nih.gov/news/imews/guidelines.htm (accessed July 5, 2002).

23. Department of Health and Human Services, news release, "Secretary Shalala Bolsters Protections for Human Research Subjects," *HHS NEWS,* May 23, 2000.

24. Department of Health and Human Services, Office of the Secretary, "Human Subject Protection and Financial Conflict of Interest: Conference," *Federal Register* 65 (July 3, 2000): 41073–41076, p. 41073.

25. At ohrp.osophs.dhhs.gov/humansubjects/finreltin/finguid.htm (accessed July 9, 2002).

26. At ohrp.osophs.dhhs.gov/humansubjects/finreltin/finguid.htm, p. 1.

27. Department of Health and Human Services, Office of Public Health and Science, Draft, "Financial Relationships and Interests in Research Involving Human Subjects: Guidance for Human Subject Protection." *Federal Register* 68 (March 31, 2003): 15456–15460.

28. Department of Health and Human Services, "Financial Relationships in Clinical Research: Issues for Institutions, Clinical Investigators, and IRBs to Consider When Dealing with Issues of Financial Interests and Human Subject Protection" in *Draft Interim Guidance* (June 10, 2001), at http://ohrp.osophs.dhhs.gov/humansubjects/finreltin/finguid.htm.

29. Lori Pressman, ed., *AUTM Licensing Survey: FY 1999* (Northbrook, Ill.: Association of University Technology Managers, 2000), 2, at www.autm.net/surveys/99/survey/99A.pdf (accessed October 18, 2002).

30. Peter Shorett, Paul Rabinow, and Paul R. Billings, "The Changing Norms of the Life Sciences," *Nature Biotechnology* 21(2003): 123–125.

31. Mildred K. Cho, Ryo Shohara, Anna Schissel, and Drummond Rennie, "Policies on Faculty Conflicts of Interest at U.S. Universities," *Journal of the American Medical Association* 284 (November 1, 2000): 2203–2208.

32. Cho et al., "Policies on Faculty Conflicts," 2208.

33. David Wickert, "UW Seldom Cuts Researcher, Corporate Ties," *The News Tribune,* October 14, 2002, A10.

34. Bernard Lo, Leslie E. Wolf, and Abiona Berkeley, "Conflicts-of-Interest Policies for Investigators in Clinical Trials," *New England Journal of Medicine* 343 (November 30, 2000): 1616–1620.

35. S. Van McCrary, Cheryl B. Anderson, Jelena Jakovljevic et al., "A National Survey of Policies on Disclosure of Conflicts of Interest in Biomedical Research," *New England Journal of Medicine* 343 (November 30, 2000): 1621–1626.

36. A. Bartlett Giamatti, "The University, Industry, and Cooperative Research," *Science* 218 (December 24, 1982): 1278–1280.

37. Giamatti, "The University, Industry, and Cooperative Research," 1280.

38. Hamilton Moses III and Joseph B. Martin, "Academic Relationships with Industry: A New Model for Biomedical Research," *Journal of the American Medical Association* 285 (February 21, 2001): 933–935.

39. Duff Wilson and David Heath, "Uninformed Consent: They Call the Place 'Mother Hutch,'" *The Seattle Times,* March 14, 2001.

40. Duff Wilson and David Heath, "He Saw the Tests As a Violation of 'Trusting, Desparate Human Beings,'" *The Seattle Times,* March 12, 2001.

13

CONCLUSION: REINVESTING IN PUBLIC-INTEREST SCIENCE

Throughout this book I have argued that the greatest danger we face from the unrestrained commercialization of our institutions of higher learning is the loss of supportive venues in which public-interest science and public intellectuals can thrive. By public-interest science, I mean the rich tapestry of involvement that professors of universities have in providing expertise to government agencies and not-for-profit organizations in addition to the pro bono help that many of them offer to underrepresented communities. Public-interest science includes the academic researcher's role as an independent voice of critical analysis on contested public issues involving his or her expertise. While the erosion of a public-interest ethos within science is the most significant loss among the other losses frequently cited—including free and open exchange of information, the knowledge commons, and the norm of disinterestedness—little has been accomplished in preventing the creeping disappearance of this critically significant contribution that universities make toward the betterment of society. Professional associations and government bodies are beginning to recognize the problem of university entrepreneurship as the primary threat to academic values. What such associations have to offer, in large part, is disclosure—the presumed universal antidote to conflicts of interest. Disclosure, however, clearly falls short of resolving the problems I have outlined; it also accepts as inevitable the abdication of the university from its traditional role as the spawning ground for public-interest science and as the reservoir of independent and disinterested expertise.

Much of the rhetoric regarding university–business linkages has focused on protecting the core values of the academic institutions—and there are certainly many reasons to do so. A society cannot pretend to stand tall among the world's great civilizations if it cannot claim a system of higher education that provides the space where scholars can create and discover without the constraints imposed on them by their institution or others external to it. In the business sector, the private control over information has a functional role; in academia, to do so is counterproductive, when all those who participate in the production and dissemination of knowledge share data, rather than hoard it.

There is much idle talk about the core values of universities, but there is little fruitful analysis on the consensus of such values. It is certainly possible to preserve academic freedom at an institution of higher learning that has long since abandoned public-interest science. The mere fact that a university holds a nonprofit status does not immunize it from such a fate. Protecting the university's core values will not, in and of itself, preserve for society the public-interest role of science (as opposed to a industry-focused role), and it will not preserve the status of the academy as the wellspring for public intellectuals (as opposed to simply another group of stakeholders).

THE UNIVERSITY'S UNIQUE STATUS

The protection of the university from becoming indistinguishable from any other business organization is not a new concern. In 1919, one of America's leading sociologists, Thorstein Veblen, author of the classic study *The Theory of the Leisure Class*, wrote a lengthy book-length commentary on the governance of universities by businessmen.[1] He saw the tendencies at the turn of the twentieth century to model higher education on what were then modern business practices, such as efficiency, quality control, labor productivity, and Taylorism (assembly line production). Veblen wrote, "If these business principles were quite free to work out their logical consequences, untroubled by any disturbing factors of an unbusinesslike nature, the outcome should be to put the pursuit of knowledge in abeyance within the university, and to substitute for that objective something for which the language hitherto lacks a designation."[2]

Business management is largely hierarchical. Veblen understood that if the university were run by businessmen exclusively in a top-down, command-control fashion, then scholars would be constrained "from following any inquiry to its logical conclusion"[3] because, he noted, some conclusions might conflict with the trustees or managers of the institution.

Of course, some form of hierarchy exists in the university, but it is a sector hierarchy primarily within the administrative arm. The academic structure is otherwise highly decentralized. Yale University's former president A. Bartlett Giamatti wrote, "the University's structure is a hierarchy unlike any other; it is neither military nor corporate; nor is it even a hierarchy like the Church whence sprang the earliest University teachers."[4] For example, neither the president nor the trustees of a university can dictate what professors teach in the classroom, what political or intellectual views they espouse, or what grants they should pursue. Writing sixty years after Veblen and having witnessed the postwar transformation of the American university, Giamatti spoke about the "ballet of mutual antagonisms" between private (for-profit) enterprise and private (not-for-profit) education, giving way to "responsible collaboration."[5]

Veblen's picture of the early twentieth-century American university was that of an institution seeking to meet its fiduciary responsibilities while supporting its teaching, scholarly, and research functions. He described its state as precarious. Was public philanthropy and tuition sufficient to protect private universities from bankruptcy? It had budgets, debts, loans, and payroll. Did it need a business plan? And was its success dependent on meeting business-like objectives according to such a plan? He wrote, "It appears, then, that the intrusion of business principles in the universities goes to weaken and retard the pursuit of learning, and therefore to defeat the ends for which a university is maintained."[6]

The university that has emerged at the dawn of the twenty-first century has adopted many principles from business management. But some of its functions and activities seem to resist such a reduction to the accountability of business norms, at least in the short-term. From Veblen's observations of nearly a hundred years ago, he could see that, in a compromise between the scholar's ideals and business principles, "the ideals of scholarship are yielding ground, in an uncertain and varying degree, before the pressure of businesslike exigencies."[7]

Derek Bok, president emeritus of Harvard, described his angst about the evolutionary trends in the modern research university toward pecuniary goals in his 1981 annual report to his overseers: "The causes of concern . . . flow from an uneasy sense that programs to exploit technological development are likely to confuse the university's central commitment to the pursuit of knowledge and learning by introducing into the very heart of the academic enterprise a new and powerful motive—the search for utility and commercial gain."[8]

What makes the American university a unique institution? What values, if any, are untouchable? In other words, throughout the thousands of local university cultures, what characteristics are conserved and recalcitrant to compromises? Where does public-interest science fit it? What are the threats to the university's core values? Among these values are the unique role of the university in American society, beyond the education of the young and its contribution to research. Giamatti noted:

> Located by charter and tradition in its independent place between the corporate part of the private sector and the public realm of government, the private university also plays another role, a role essential in a pluralistic, free society. And that

> is the role of independent critic, critic of the private sector it inhabits, of the government it respects, of itself as an institution and as the guardian of a process.[9]

What are the features and characteristics of the university that make it distinctive? What contributions to society will be lost when it is captured under the entrepreneurial umbrella? At the research universities, faculty have three roles: teaching and advising; scholarship or scientific research; and administration, such as service on committees. The academic arm of the university is largely self-directed. Departments define and teach the courses of their choosing, following the guidelines established by their disciplines. They are accountable only to the extent that they stray too far afield from the canons of a discipline, especially when they are undergoing accreditation by their respective accrediting organizations.

Administrators can show frustration with the independence of tenured faculty. A provost at Tufts University was often heard pining that running a faculty meeting was like herding cats. James Duderstadt, president emeritus of the University of Michigan wrote that many universities possess a structure that "can best be characterized as creative anarchy, in which faculty leaders in posts such as department chair simply do not have the authority to manage, much less lead their units . . . the university remains very much a bottom-up-organization, a voluntary enterprise."[10] Former dean at Harvard, Henry Rosovsky was quoted as saying that "the faculty has become a society largely without rules, or to put it slightly differently, the tenured members of the faculty, frequently as individuals, make their own rules."[11]

Faculty are responsible for teaching their courses, which at the major research universities range from three to nine class hours per week for about twenty-eight weeks a year. Professors in the natural sciences who supervise laboratories teach, on average, less than the professors in the humanities and social sciences.

One of the distinctive aspects of university life for a faculty member is that most of his or her time is self-directed. For teaching, this discretion means that faculty choose the courses they teach and their content. Of course, many universities have required courses as part of a general education's core curriculum, and departments naturally have required courses for a major. For example, statistics is required in most schools for psychology majors. There are some basic concepts and methods that are covered by nearly all the beginning statistics courses, just as there are beginning physics and calculus classes that likewise provide the core knowledge that is central to these fields. In addition to the required classes are the standard courses that must be taught in certain disciplines, such as microeconomics in departments of economics. Beyond the core courses for a discipline, however, professors have enormous flexibility within academia in experimenting with new courses, including those that encompass cross-disciplinary subject matter or methods of analysis.

Even after they receive tenure, faculty are expected to publish and do research as part of their permanent job designation. But the soul of the university is premised on the idea that academic faculties are free to think, write, research, and investigate what they choose. For example, someone can begin a career as a botanist studying exotic plant species in Brazil. But if he chooses, he may turn his attention to the study of rare fungi in Africa. If he continues to be productive, he will be compensated with raises and promotion. If he is no longer productive as a scientist–scholar, he may be asked to take on more teaching and administrative responsibilities.

At leading American universities, faculty by and large have considerable time beyond their research, teaching, and administrative roles to devote to public service activities. Faculty members are usually given one day per week to consult or do personally rewarding work that may not be related to their university functions.

American university scholars lend their names to all sorts of causes and serve on the boards of public interest and professional organizations. For example, I am a founding board member of the Council for Responsible Genetics, and I provide pro bono services to the organization.

Universities cannot prohibit faculty from using their university and department affiliation when they express views, expert or otherwise, on issues related or unrelated to their area of expertise. The structure of American universities are set up to support the role of public intellectuals. Experts in many fields serve a greater public function, beyond their scholarly research, by speaking directly to citizens through writings in popular magazines, commentaries, op-ed essays, and through media interviews. This sort of attention is not comfortable to all faculties. Some shy away from the public eye. But the vast numbers who do participate in civic dialogue do so with the full recognition that they are received as independent voices—that is, not under the influence of their institution. They are neither encumbered nor controlled by their institutional affiliation.

When the Black studies and religion scholar Cornell West was rebuked by Harvard President Lawrence Sommers for participating in the creation of a compact disc on Black urban culture, the overall reaction of the academic community was swift and highly critical of Sommers. He stepped over the line by interfering with Professor West's academic freedom to decide how he wishes to devote his creative energy, with what issues, and with whom he chose to associate and lend his name.

All citizens, of course, share the right of free expression and association. But the reality is that most people working in fields where their public voice would have special significance, such as government regulators or scientists working in industry, face explicit institutional norms or implicit taboos against public expression on issues that affect their organization. The exception can be found with the nonprofit advocacy organizations, where speaking forcefully on issues is part of the job description. Nevertheless, it is assumed that those who work for these groups will be faithful to the organization's political perspective; otherwise, they will lose their job. Thus, a group founded to alert citizens about the dangers of global warming will not tolerate an employee who speaks publicly in support of new incentives for burning coal.

At universities, there is no institutional consensus on any issue. The concept of academic freedom implies that faculty members are expected to speak their conscience or their commitment based on what they have learned. For this reason, the government, the media, and the legal community depend on experts spread throughout the academic landscape to provide an informed analysis of an issue. Some faculty even re-create themselves as experts in fields for which they were not hired or in which they are not formally educated. Once they have proven their value to society, they too are pursued for their independent viewpoints.

When there is reason to believe that a university as an institution, or some division of the university, has a conflict of interest, the trust in its faculty gets eviscerated rather quickly. As an example, when scientists from the University of California, Berkeley, published findings that the traditional corn varieties in southern Mexico were contaminated by genetically modified varieties of corn (see chapter 3), their colleagues in another department criticized their results. However, some members of the media were skeptical of the critics whose department, en masse, had signed a contract with one of the nation's leading biotechnology corporations. In the words of Giamatti, "The university is an independent institution in our society, and that it cannot serve society responsibly unless that independence is its paramount concern."[12]

SOCIAL VALUE OF ACADEMIC FREEDOM

One of the features distinguishing the university from other institutions, private and public, in European and American society is "academic freedom." It is the sine qua non of academic life. In their book *The American University*, Parsons and Platt wrote that academic freedom "is the right to conduct cognitive exploration and communication with minimal preimposed constraints" and that it "includes the right to express and advocate extreme views."[13] The legal concept of academic freedom has been traced back to Germany around

1850 when the Prussian Constitution asserted the freedom of scientific research (*Freiheit der Wissenschaft*).[14]

The concept of "protected employment" of professors from external affairs has its roots in the Middle Ages, when medieval scholars "sought to be autonomous and self-regulating to protect knowledge and truth from corrupt outside influences."[15] In 1650, Harvard University was chartered; in 1722, the university appointed Edward Wigglesworth its first professorship (of Divinity) without limit of time, thus ushering in the tenure system to North America. Two hundred years later, in 1925, the fledgling American Association of University Professors (AAUP) addressed the issue of tenure in connection with academic freedom at its first national conference. The contemporary view of tenure in higher education, however, can be traced to the 1940 AAUP *Statement on Principles of Academic Freedom and Tenure.* By the late 1950s, the lifetime appointment of university faculty at American universities after a successful probationary period had become almost universally accepted in higher education. The AAUP position on tenure has since been endorsed by over 170 professional societies, including the Association of Colleges and Universities and the Association of American Law Schools. Tenure in higher education, which initially developed out of political and economic motivations, became inextricably linked to academic freedom.[16] Aronowitz wrote in *The Knowledge Factory* that "we associate the institution of tenure with the need to protect dissenting faculty from sanction imposed by the public, the administration of the university, and colleagues who might be prone to punish apostates."[17]

Those who understand that they cannot be dismissed from employment for any cause (except for nonperformance of their required duties as a teacher or for gross moral infractions) will have the security, at least in theory, that they can speak independently on any issue. The burden of proof to dismiss a tenured faculty member is and should be exceedingly high. Contrary to popular thought, tenure is not an absolute guarantee of lifetime employment. For the most part, universities and the courts have supported the principles behind tenure. Jacques Barzun wrote, "The practice of giving tenure to university professors is justified on the same grounds as the tenure of judges in the federal courts: to make them independent of thought control."[18]

Presidents of universities typically are appointed by the Boards of Trustees as "at-will employees," which means that they can be discharged by the Board without being provided with a cause or without being given due process. University presidents have no guarantees of administrative tenure and, for many, no guarantee of a specified number of employment years. Only the faculty has tenure. Faculty members who are appointed to the presidency of their university, however, retain tenure in their discipline, even when they step down from the presidency. If a tenured faculty member is appointed to a high administrative post at another university, then that appointment does not automatically come with tenure (unless they are retenured in a specific department). University presidents, who typically do not have the job security of faculty, are increasingly finding it more difficult than they had in the past to speak out freely on public issues. Taking a strong moral stand on a contemporary public issue might create controversy and thus adversely affect funding sources among foundations, alumni, state legislatures, or big donors, who are generally from the corporate sectors.[19] This reluctance to speak may explain the fact that in recent years few university presidents have risen to the stature of public intellectuals in comparison to presidents of years past.

Two conditions exist under which academic freedom may become extinct at universities. The first condition is the eradication of tenure as we know it. There is no better protection of the rights of faculty to teach, write, and undertake research of their choosing than the award of tenure. It is well understood among junior faculty that their probation period is not a time to test the limits of academic freedom. Junior professors are far more cautious about their public pronouncements, their political affiliations, and their involvement in campus and national politics.

Freedom of inquiry and the expression afforded to faculty must be more than an ideal or a principle. They must be protected concretely against even the remote possibility that bias against one's beliefs can contaminate decisions of administrators who hire and fire. There certainly could be legal protections for teachers and scholars from being unjustly dismissed, even where tenure is not awarded. But the burden of proof in such cases would fall to the faculty member, who would have to take legal action against a far wealthier opponent. Nothing less than tenure establishes the high burden of proof needed to protect a senior professor from dismissal against allegations that would surely be surrogates for cleansing the university of certain unpopular ideas. As tragic as it sounds, once tenured, a professor may turn into an unenlightened bigot and still have job security. The protection of academic freedom in the rare and unpopular cases is ultimately what establishes the university as a unique social institution.

The second condition that could undermine academic freedom is much more subtle. The possibility exists that universities will not tamper with academic freedom explicitly and directly. But if the university changes its form, its dependency for funding, and its ethical norms, its professors will find it uncomfortable or counterproductive to realize the opportunities afforded by academic freedom. It will become an abstract right analogous to voting in a society with high poll taxes or to voting for candidates who are so much alike that citizens do not feel that their vote can make a difference. What is the value of having academic freedom when professors are suspended between two cultures, one of which is outside the university, where serving as a public critic is considered a taboo? The social value of academic freedom is its robust use. Overlapping cultures of universities and business can turn the role of academic freedom into an anachronism.

A professor who is involved in a start-up company that is developing one of his patented inventions enters a subculture of venture capital, advertising, marketing, stocks, loans, regulations, and all the trimmings of entrepreneurship. Within this subculture, certain norms of behavior are quite distinctive of those generally found at the university. "Boldness" means taking business risks; it does not mean criticizing a public policy. "Adventurous" means marketing an unusual product; it does not mean putting forth a new theory of disease causation. Academic entrepreneurs behave differently than public intellectuals. The former markets products while the latter promotes ideas. In business circles, it is uncommon for a CEO of one company to speak publicly about dangerous products or occupational hazards. Corporate executives are increasingly expected to operate within the contested zone of good business practice and dubious ethical behavior. There is no clear boundary, only a blurred one.

For academics, criticism is essential to the growth of knowledge, and public criticism is a key to finding the truth. It is not only proper, but also obligatory, to criticize another's ideas. Thus, academics who mix business with the pursuit of knowledge will probably internalize the norms of the new subculture resulting in their avoidance of public criticism. As an academic entrepreneur, it is expected that you accept some of the basic rules of entrepreneurship. How likely is it that someone who has a patent on a gene would speak out against "life patents"? It would not even enter that person's consciousness.

An example that illustrates the near corporate capture of a field of academic science can be found in weed science. In their book *Toxic Deception*, Fagin and Lavelle write, "Weed scientists—a close knit fraternity of researchers in industry, academia, and government—like to call themselves 'nozzleheads' or 'spray and pray guys.'"[20] Because federal grants are sparse, university weed scientists depend almost entirely on industry support. As a result, their research focuses primarily on chemical herbicides rather than on alternative forms of weed management, such as crop rotation strategies and biocontrol mechanisms. The authors quote a USDA agricultural scientist and past president of the Weed Science Society of America: "If you don't have any research other than what's coming from the ag chemical companies, you're going to be doing research on agricultural chemicals. That's the hard cold fact."[21] In 1996, a colleague and I published a study on agricultural biotechnology where we surveyed the articles in the journal *Weed Science* between 1983 and 1992 to determine how the published work reflected different lines of research. We found that about 70 percent of the research focused on chemical herbicides and 20 percent on weed biology and ecology, with less than 10 percent devoted to integrated pest management and nonchemical approaches to weed control.[22] If a subfield of science is largely supported by industry, then it scarcely matters whether scientists are working in corporate facilities or university facilities. The problems that are framed, the type of data collected, and the effects measured are largely under the influence of the industry sponsor. Under these circumstances it would not be considered proper or wise for scientists so beholden to one segment of the private sector to engage in public-interest science by questioning the strategy used to manage weeds.

Campbell and Slaughter studied the values and normative behavior of scientists involved in university–industry relationships. In a superb piece of government-supported research, they found that the university scientists begin to accommodate to the norms of their corporate partners. As a result, they reported that the "opportunities for inappropriate actions are likely to increase." They went further to state that this new relationship "creates a climate that multiplies the opportunities for misconduct."[23] Until their study, no analyst of university–industry relations was confident enough to draw any connections between industry partnerships and scientific misconduct.

The science professorate, once a calling for scholar–teachers, has become a staging ground for self-interested entrepreneurs who want the dignity and prestige of the position and the freedom to advance their pursuit of personal wealth. University of Michigan's President James Duderstadt commented that "the increasing specialization of faculty, the pressure of the marketplace for their skills, and the degree to which the university has become simply a way station for faculty careers have destroyed institutional loyalty and stimulated more of a 'what's in it for me' attitude on the part of many faculty members."[24]

Public-interest science requires people who are able and willing to speak out candidly and critically about toxic dangers, political injustice, pernicious ideology that foments hatred, false idols, unsustainable paths of economic development, product liability, and environmental causes of disease. Speaking about the value of private universities in society, Giamatti wrote that they play an essential role in a pluralistic society of "independent critic, critic of the private sector it inhabits, of the government it respects, of itself as an institution and as the guardian of a process."[25] Of course, the university as critic is the sum total of its faculty who choose to exercise academic freedom for the benefit of society. The concept of the university's service to society is highlighted in the statement by the Association of Governing Boards of Universities and Colleges: "Public and independent nonprofit colleges and universities are unique among social institutions in that their mission require them to work to benefit the whole of society through teaching research and service. . . . The [universities] constitute a precious reservoir of expertise and cultural memory that simultaneously serves the past, the present and the future."[26] How will this role be protected?

With almost predictable periodicity, university presidents and academic boards of trustees meet to discuss a perennial issue: how to protect the cherished values of the university from erosion, compromise, or transgression. Such a meeting was held in Pajaro Dunes, California, in the spring of 1982 at the dawn of the new era of university–industry collaborations. The presidents of Stanford, Harvard, MIT, Cal Tech, and the University of California hosted three dozen conferees from universities and industry to discuss the future of university–industry relations and the preservation and integrity of the university and its faculty. At such meetings, one can often hear the expressions of lofty ideals, such as "academic freedom," "free and open exchange of knowledge," or the university as a "credible and impartial resource." But one also hears that university–industry partnerships, technology-transfer entrepreneurship, and faculty-based companies are a reality of the modern university. The report from Pajaro Dunes recommended that universities continue to support the tradition of protecting free inquiry while supporting the new arrangements to attract corporate partnerships.[27]

Traditionalists want to hold the line against crass commercialization. The modernists argue that the boundaries of American institutions are shifting and that universities must accommodate to these shifts. The tragic death of Jesse Gelsinger, the result of a failed gene therapy experiment, accented the seriousness of conflicts of interest in clinical research and drew a strong response from the federal government. The DHHS commissioned the Institute of Medicine (IOM) of the National Academy of Sciences to perform a comprehensive assessment of the regulations for the protection of human subjects. In 2002, the panel convened by IOM published its findings, which included some far-reaching changes in the national system for the protection of research risks. The panel issued a strong recommendation to elevate the importance of evaluating conflict of interest in the review of protocols for clinical trials. In its report, the panel suggested three distinctive reviews by different individuals or boards for each human-subjects protocol: a review of the science; a review of the financial conflicts of interest; and a review of the other ethical issues associated with the study, such as acceptable risks and informed consent. Under the IOM proposal, a new research Ethics Review Board (ERB) would replace the traditional Institutional Review Board (IRB) that has been a mandated part of the federal regulations since the 1970s. Once the science and financial conflict of interest reviews are completed, according to the proposal, the research ERB would make a final determination on the ethical acceptability of the protocol.

This is the first major effort by the National Academy of Sciences and its Institute of Medicine to establish conflict of interest as a primary component in the federal system for protecting human subjects. The panel report noted, "Potential conflicts of interest of the investigator, Research ERB members, or the institution should be assessed by the organization's relevant conflict of interest oversight mechanism and communicated to the Research ERB. . . . The conflict of interest oversight body should determine whether financial conflicts should be disclosed, managed, or are so great that they compromise the safety or integrity of the proposed research."[28]

If reformed according to the IOM recommendations, human-subject protection will be improved by making conflicts of interest more transparent. But nothing has yet been proposed to decrease or minimize the financial conflicts of interest that occur in publicly funded universities and nonprofit research institutes.

REESTABLISHING TRADITIONAL ROLES AND BOUNDARIES

We are in an era where traditional sector boundaries are disappearing. Banks have brokerage businesses; brokerages do banking. The entertainment industry owns news broadcasting; news broadcasters have taken to entertainment formats. Farms are used to produce agrichemicals; industrial fermenters are used to manufacture food. Animals are being used to make drugs; drugs are used for entertainment. Government researchers are permitted to form partnerships with companies; companies contribute money to federal research on toxic chemicals. Independent auditing firms, responsible for generating federally mandated accounting data on public corporations, provide other consultancies to firms they audit. A pharmaceutical manufacturer of oncology drugs takes over the management of a cancer treatment center. The world is a little topsy-turvy. Perhaps we shouldn't be surprised to learn that nonprofit institutions, like universities, are engaging in aggressive for-profit ventures. It is another example of the convoluted boundaries of institutions.

What can be done to regain the integrity of academic science and medicine at a time when turning corporate and blurring the boundaries between nonprofit and for-profit are in such favor? We should perhaps begin by reexamining the principles on which universities are founded. We should then reexamine the importance of protecting those principles from erosion and compromise for the sake of amassing larger budgets and providing more earning potential for select members of the faculty. Three principles should guide our approach:

- The roles of those who produce knowledge in academia and those stakeholders who have a financial interest in that knowledge should be kept separate and distinct.
- The roles of those who have a fiduciary responsibility to care for patients while enlisting them as research subjects and those who have a financial stake in the specific pharmaceuticals, therapies, products, clinical trials, or facilities contributing to patient care should be kept separate and distinct.
- The roles of those who assess therapies, drugs, toxic substances, or consumer products and those who have a financial stake in the success or failure of those products should be kept separate and distinct.

The public's trust in the integrity of academic science and medicine begins to eviscerate when these roles are conflated. The first operational test toward achieving these goals is the disclosure of one's financial interests. The disclosure, by scientists, of financial stakes in research is but the first step in satisfying the general principles because such disclosure communicates whether a general principle is likely to be violated. The declaration of financial interests by university scientists should be universally required by journals, congressional committees, granting agencies, and federal advisory panels. Conflict-of-interest disclosures should be as much a part of grant submissions as it is a part of disseminating the results of research. Federal granting agencies can create incentives for journals to adopt COI policies by requiring grantees to disclose any financial conflicts of interest when they submit articles for publication that arose from the research funds.

Disclosure of interests does not satisfy the principles unless efforts are made to prohibit certain relationships. Faculty members who are principals or officers in a company or who have substantial equity interest in a company should be prohibited from engaging in research sponsored by that company. Nothing less will ensure that a firewall exists between a principal stakeholder of the knowledge and the investigator.

Academic institutions that are major stockholders in companies (i.e., they hold sufficient stock to enable them to influence management decisions) beyond what is considered the norm for an investment portfolio of a nonprofit university should recuse themselves from accepting any research grants or contracts sponsored by that company. Alternatively, if they do wish to engage in such research, they should establish a blind trust for the stock that could be affected by the outcome of the research.[29] Only under these conditions (or something equivalent to them) will research and investment decisions be appropriately disengaged from one another.

A similar arrangement applies to the health sciences. A physician who supervises clinical trials must have no equity interests in a company that stands to gain or lose from the outcome of the trials. The fiduciary responsibility of the clinical investigator must be to the integrity of the science and to the well-being of the human subjects. These goals may, at times, be in conflict, but they are not severable. However, to mitigate potential conflicts of interest that arise in the fiduciary responsibilities of clinical investigators, physicians who have no affiliation with the clinical trial should be available as patient advocates. The clinical investigator will therefore have another medical opinion, unrelated to the trial, on such questions as: Should the patient be allowed to continue in the trial? Should the patient continue to receive the placebo?

The protection of federal advisory panels against the undue influence of stakeholder groups is essential to sound public decision making. Scientific expertise should not be decided by a political litmus test.[30] Government bodies should take a literal interpretation of federal statutes that exclude conflicts of interest on many science-based advisory committees. Waivers of candidates for advisory committees who have significant financial interests in the subject matter of the panel should meet a high burden of proof. And when, in rare circumstances, individuals meet this burden, they should serve as ad hoc, nonvoting members on the advisory panel. Their conflicts of interest should be a matter of public record.

We need strong federal incentives for rewarding academic scientists for maintaining their independence from entrepreneurial activities when they are appointed to prestigious panels or when they are awarded grants and contracts; otherwise, the process of establishing an honest system of science-based policy will fall into a vicious circularity. Government agencies argue that the best experts in the field have a conflict of interest. Such experts get appointed to the prestigious panels and thus receive an ample number of grants; such grants raise the experts' stature in the scientific community; such stature makes them even more sought after on the expert panels. All of this leads to more conflict-of-interest waivers and helps to normalize stakeholder interests in government panels. The cycle can only be broken when the incentive structure for grants, contracts, and panel appointments does not reward the multivested interests of entrepreneurial scientists and when it makes the waiver of such interests the exception rather than the rule.

Among the most challenging and serious problems of conflict of interest pertains to the role of academic scientists in the assessment of drugs, toxic chemicals, and other consumer products. In drug testing, for example, companies typically contract out the studies, many of which go to university researchers in fields such as medicine, toxicology, and pharmacology. The sponsoring companies often have control over the data and the publications; they help design the protocols; they then submit the results to the relevant agency, such as the FDA. Similar practices are found in testing chemicals where data get submitted to the EPA or to OSHA (Occupational Safety and Health Administration).

The close connection between the sponsors of product assessments and academic researchers is unsettling. Some institutions and investigators may be willing to compromise the rigor, integrity, and independence of the study in exchange for the assurance that more funding will follow. In theory, nothing can prevent this scenario from happening. The best approach would be to establish a firewall between those who assess and those who would benefit from a particular outcome. It is impractical to imagine, however, a government bureaucracy large enough to undertake such testing.

A degree of imagination will be necessary to protect the integrity of product assessment research paid for and (to some degree) controlled by stakeholders. One approach, say for drug testing, would be to establish an independent national institute for drug testing (NIDT). Any company wishing to generate data for submission to the FDA would have to submit the drug to the NIDT. The drug-testing institute, in collaboration with the private sponsor, would issue a drug-testing protocol. The NIDT would put out a notice to research centers, called a request for proposal (RFP). Qualified drug assessment centers would submit their proposals, and the NIDT would select a testing group. Protocols, data utilization, and publications would be negotiated between the institute and the contractee. The institute would have quality-control requirements that would be applied uniformly. Proprietary information would be protected. Once the test is completed, the results would be sent to the company. As certified data, it can thus be submitted to the relevant agency.

In this scenario, the company is still permitted to use any outside testing firm for all sorts of preliminary findings. But only data coming from the NIDT can be used to gain drug approval. The NIDT would compile documentation on all studies done through the institute, whether they show positive or negative results on the tested product. This type of framework would require a new government agency to create a buffer zone between the stakeholders and the drug testers, thereby protecting all parties from potential conflicts of interest. The possibilities for bias would be significantly diminished. Conflict-of-interest rules implemented by the NIDT would be applied to testing groups whether they are situated at universities or at private firms. No tester or testing institution could have equity in a company poised to benefit from the testing results.

I have argued that the stakes of introducing excessive commercialization into the university has impacts well beyond the halls of academia or the professional scientific associations. Each publicly chartered university in the United States plays a unique social role in the stewardship of a large enterprise called "the production of knowledge." No single university bears this responsibility; rather, it is the aggregate of America's research universities. This intellectual wellspring is where most of this country's journal editors come from. It is where the presidents of professional scientific societies are chosen. It is where we find most of our candidates for the National Academy of Sciences. And it is the source of the vast preponderance of original research.

If we contaminate the wellspring of knowledge by mixing other interests, in particular corporate agendas, then we lose the pure reservoir for dispassionate and independent critical analysis. Individual universities, because of their institutional conflicts of interest, will be seen as another stakeholder—another self-interested party in a cynical political arena where truth is all too often seen as a social construct and not as an objective outcome of rigorous scholarly investigation.

Certain institutions must be protected by tradition, law, or regulation from taking on conflicting roles. The court system is not the same institution that runs the prisons. Physicians should not be earning income every time a person swallows a pill or participates in a clinical trial. Members of Congress and judges should not be sitting on the boards of corporations. University scientists should not be corporate CEOs or handmaidens to America's for-profit companies. By accepting the premise that conflicts of interest in universities must be subtly managed, rather than prohibited or prevented, nothing less than the public interest function of the American academic enterprise is at stake.

NOTES

1. Thorstein Veblen, *The Higher Learning in America* (New York: B. W. Huebsch, 1919).
2. Veblen, *The Higher Learning in America,* 170–171.
3. Veblen, *The Higher Learning in America,* 186.
4. A. Bartlett Giamatti, *The University and the Public Interest* (New York: Athenium, 1981), 18.
5. Giamatti, *The University and the Public Interest,* 23.
6. Veblen, *The Higher Learning in America,* 224.
7. Veblen, *The Higher Learning in America,* 190.
8. Barbara L. Culliton, "The Academic–Industrial Complex," *Science* 216 (May 28, 1982): 960–962.

9. Giamatti, *The University and the Public Interest,* 114.

10. James J. Duderstadt, *A University for the 21st Century* (Ann Arbor: University of Michigan Press, 2000), 163–164.

11. Cited in Duderstadt, *A University for the 21st Century,* 154.

12. Giamatti, *The University and the Public Interest,* 132.

13. Talcott Parsons and Gerald M. Platt, *The American University* (Cambridge, Mass.: Harvard University Press, 1975), 293.

14. Ronald B. Standler, *Academic Freedom in the USA,* at www.rbs2.com/afree.htm (accessed June 18, 2002).

15. At www.uh.edu/fsTITF/history.html (accessed June 18, 2002).

16. Richard Hofstader and Walter P. Metzger, *The Development of Academic Freedom in the United States* (New York: Columbia, 1955).

17. Stanley Aronowitz, *The Knowledge Factory* (Boston: Beacon Press, 2000), 65.

18. Jacques Barzun, *The American University* (New York: Harper and Row, 1968), 59.

19. This was pointed out to me by John DiBiaggio, who held the presidency at three universities: University of Connecticut, Michigan State, and most recently, Tufts University.

20. Dan Fagin and Marianne Lavell, *Toxic Deception* (Secaucus, N.J.: Carol Publishing Group, 1996), 52.

21. Fagin and Lavell, *Toxic Deception*, 53.

22. Sheldon Krimsky, *Agricultural Biotechnology and the Environment* (Urbana: University of Illinois Press, 1996), 52–53.

23. Teresa Isabelle Daza Campbell and Sheila Slaughter, "Understanding the Potential for Misconduct in University-Industry Relationships: An Empirical View," in *Perspectives on Scholarly Misconduct in the Sciences,* ed. John M. Braxton (Columbus: Ohio State University Press, 1999), 259–282.

24. Duderstadt, *A University for the 21st Century,* 163.

25. Giamatti, *The University and the Public Interest,* 227.

26. Association of Governing Boards of Universities and Colleges, "AGB Statement on Institutional Governance and Governing in the Public Trust: External Influences on Colleges and Universities," Washington, D.C., 2001.

27. Ann S. Jennings and Suzanne E. Tomkies, "An Overlooked Site of Trade Secret and Other Intellectual Property Leaks: Academia," *Texas Intellectual Property Law Journal* 8 (2000): 241–264.

28. Institute of Medicine, Committee on Assessing the System for Protecting Human Research Participants, *Responsible Research: A Systems Approach to Protecting Research Participants* (Washington, D.C.: The National Academies Press, 2003), 74.

29. Hamilton Moses III and Joseph B. Martin, "Academic Relationships with Industry: A New Model for Biomedical Research," *Journal of the American Medical Association* 285 (February 21, 2001): 933–935.

30. David Michaels, Eula Bingham, Les Boden et al., "Advice without Dissent," editorial, *Science* 258 (October 25, 2002): 703.

SELECTED BOOKS

Bok, Derek. *Universities in the Marketplace: The Commercialization of Higher Education.* Princeton, N.J.: Princeton University Press, 2003.

Greenberg, Daniel S. *Science, Money and Politics: Political Triumph and Ethical Erosion.* Chicago: The University of Chicago Press, 2001.

Kenney, Martin. *Biotechnology: The University-Industrial Complex.* New Haven: Yale University Press, 1986.

Porter, Roger J., and Thomas E. Malone. *Biomedical Research: Collaboration and Conflict of Interest.* Baltimore: The Johns Hopkins University Press, 1992.

Rodwin, Marc A. *Medicine, Money & Morals: Physicians' Conflicts of Interest.* New York: Oxford University Press, 1993.

Slaughter, Sheila, and Larry L. Leslie. *Academic Capitalism: Politics, Policies and the Entrepreneurial University.* Baltimore: The Johns Hopkins University Press, 1997.

Solely, Lawrence C. *Leasing the Ivory Tower: The Corporate Takeover of Academia.* Boston: South End Press, 1995.

Weisbrod, Burton A., ed. *To Profit or Not to Profit: The Commercial Transformation of the Non-Profit Sector.* Cambridge, U.K.: Cambridge University Press, 1998.

Yoxen, Edward. *The Gene Business.* New York: Harper & Row, 1983.

INDEX

ABOUT THE AUTHOR

Sheldon Krimsky is professor in the Department of Urban and Environmental Policy and Planning at Tufts University's School of Arts and Sciences and adjunct professor in the Department of Family Medicine and Community Health at the Tufts University School of Medicine. He holds B.S. and M.S. degrees in physics from Brooklyn College and Purdue University and an M.A. and Ph.D in philosophy from Boston University. His writings have focused on the interface of science, ethics, and policy. Professor Krimsky is the author of more than 130 essays and reviews, and author, coauthor, and coeditor of six other books.

Professor Krimsky served on the National Institutes of Health's Recombinant DNA Advisory Committee. He was a consultant to the Presidential Commission for the Study of Ethical Problems in Medicine and Biomedical and Behavioral Research and to the Congressional Office of Technology Assessment. He also served as chair of the Committee on Scientific Freedom and Responsibility for the American Association for the Advancement of Science. Professor Krimsky was elected Fellow of the American Association for the Advancement of Science for "seminal scholarship exploring the normative dimensions and moral implications of science in its social context."

OTHER BOOKS BY SHELDON KRIMSKY

Genetic Alchemy: The Social History of the Recombinant DNA Controversy (1982)

Environmental Hazards: Communicating Risks as a Social Process (with A. Plough) (1988)

Biotechnics and Society: The Rise of Industrial Genetics (1991)

Social Theories of Risk, ed. (with D. Golding) (1992)

Agricultural Biotechnology and the Environment (with R. Wrubel) (1996)

Hormonal Chaos: The Scientific and Social Origins of the Environmental Endocrine Hypothesis (2000)